Lower Turonian inoceramids from Sergipe, Brazil: systematics, stratigraphy and palaeoecology

MARIA HELENA RIBEIRO HESSEL

Project 242
CRETACEOUS OF
LATIN AMERICA

Hessel, Maria Helena R. 1988 10 31: Lower Turonian inoceramids from Sergipe, Brazil: systematics, stratigraphy and palaeoecology, *Fossils and Strata*, No. 22, pp. 1–49. Oslo. ISSN 0300-9491. ISBN 82-00-37414-9. [Revised and published version of doctoral thesis presented at the Faculty of Science, Uppsala University, 1987.]

Lower Turonian inoceramid bivalves are described from the Sergipe Basin in northeastern Brazil, and a palaeoecological analysis of the region is attempted. The principal locality, Retiro 26, comprises a sequence of 35 m of the Cotinguiba Limestone Formation. Various unconformities occur in the sequence. The formation is inferred to have been deposited in an outer shelf environment of a subtropical shallow sea, with oscillating water depth, occasional strong bottom currents and ephemeral reductions of the oxygen level. The inoceramids show some endemism. Two inoceramid associations are recognized: the *Mytiloides mytiloides* and the *Mytiloides hercynicus* associations. Four new species of divergently ornamented inoceramids are described and referred to the new genus *Rhyssomytiloides*. Two new species of *Sergipia* and some previously known species of *Mytiloides* and *Rhyssomytiloides* are also described. □ *Inoceramidae, lower Turonian, Upper Cretaceous, Nordeste, northeastern Brazil, Sergipe Basin, Cotinguiba Formation, South Atlantic Ocean, taxonomy, palaeoecology, biostratigraphy,* Rhyssomytiloides, Mytiloides, Sergipia, *new taxa, new localities.*

Maria Helena Ribeiro Hessel, Paleontologiska institutionen, Box 558, S-751 22 Uppsala, Sweden; present address: Departamento de Geociências, Universidade de Brasília, C.P. 152775, 70919 Brasília, DF, Brazil; 1986 06 01 (revised 1987 11 26).

Contents

Introduction

Following the studies of Reyment & Tait (1972) on the development of the South Atlantic during the Cretaceous, Peter Bengtson carried out field-work in the Sergipe Basin of northeastern Brazil in 1971–1972 and 1977. I had the opportunity of participating in the 1977 investigations and since then have studied the lower Turonian inoceramids of Sergipe. I revisited the area in June 1981, March 1982 and February 1983. The inoceramids collected by Bengtson in 1971–1972 from the Cenomanian–Coniacian Cotinguiba Formation are being studied by Erle G. Kauffman of the University of Colorado, U.S.A.

The present work is based on a detailed stratigraphical survey of a 35 m limestone sequence exposed in the quarry Retiro 26, belonging to the lower Turonian, through the systematic collection of approximately 150 specimens of inoceramids. The area adjacent to this locality was studied to provide additional information on the sedimentary sequence. Three other localities (São Roque 5, Retiro 17 and Retiro 21 of Bengtson 1983) were also sampled, as they yielded the same forms of previously unknown, divergently ornamented inoceramids as Retiro 26.

The inoceramids of the Retiro 26 succession are described here along with a discussion of the lithological and palaeontological details. A series of environmental reconstructions is presented and the palaeogeography and biostratigraphy are discussed in the final section.

The results presented here give previously unknown details of the varied inoceramids found in the Cretaceous of Sergipe, which contains one of the richest faunas in the Southern Hemisphere. It should contribute to a deeper understanding of the inoceramid biostratigraphy and of the poorly known Cretaceous bivalve faunas of South America.

Acknowledgments. – This paper is a slightly revised version of my doctoral dissertation presented at Uppsala University in November 1987. I am deeply grateful to Dr Peter Bengtson who through his constant assistance, patience and understanding, together with his keen critical judgment, has followed and guided me through the elaboration of this study. It was he who inspired me to start studying the inoceramids of Brazil during my first participation in field-work in Sergipe.

I would also like to thank geologist Albertino de Souza Carvalho and Ulf Gregor Baranow for all the work, dedication and companionship during the field-work.

I express my gratitude to the Companhia de Cimento Portland de Sergipe of the Votorantim Group, for allowing the collection of specimens from their quarry, and to the geologists Paulo Ianez V. de Lima, Jacinto Alves de Carvalho Neto, and Manoel Henrique F. Neto, who helped in many ways with my work at Retiro 26.

Dr Jacques Sornay (Tain de l'Hermitage) kindly showed his collections of Turonian inoceramids and discussed their taxonomy with me. Drafts of the manuscript were kindly read by Drs Gerhard Beurlen (Petrobrás, Rio de Janeiro), Othon Henry Leonardos (University of Brasília), Björn Malmgren (Uppsala University), Tatsuro Matsumoto (Kyushu University), Richard A. Reyment (Uppsala University), Enrico Savazzi (Uppsala University) and Jacques Sornay, who made valuable and critical suggestions.

My thanks go also to Dr Annie V. Dhondt (Brussels) who, as my thesis examiner, helped me to eliminate a number of errors in this publication.

My gratitude is due to palaeontologists Diógenes de Almeida Campos and Geraldo da Costa Barros Muniz for making available the inoceramids kept at the Departamento Nacional da Produção Mineral (DNPM) and the Universidade Federal de Pernambuco, respectively.

I am obliged to Gustav Andersson and Tommy Westberg for the photographs of the inoceramids at Uppsala University, to Martin Feuer for additional darkroom work, and to Suzana Bengtson for guidance in the casting and the preparation of the specimens. All illustrations, except for the photographs of the inoceramids, are by my own hand.

Financial support was received from the Coordenação de Aperfeiçoamento de Pessoal de Nível Superior (CAPES) of the Brazilian Ministry of Education, the Conselho Nacional do Desenvolvimento Científico e Tecnológico (CNPq) of the Brazilian Ministry of Science and Technology, the Swedish Institute, and the Department of Palaeontology at Uppsala University.

I would also like to thank Drs Amadeu Cury, Darcy Closs, Othon Henry Leonardos and Ulf Gregor Baranow for their constant encouragement, without which this thesis would not have been accomplished. For their invaluable logistics support I want to express my most sincere gratitude to the Carneiro and Bengtson families in Uppsala. To all who indirectly contributed to the fullfilment of this work I would like to state here my gratitude.

The Turonian of the Sergipe Basin

Geological outline

The Sergipe–Alagoas Basin is one of the several basins along the Brazilian coast. It is situated in the eastern part of the

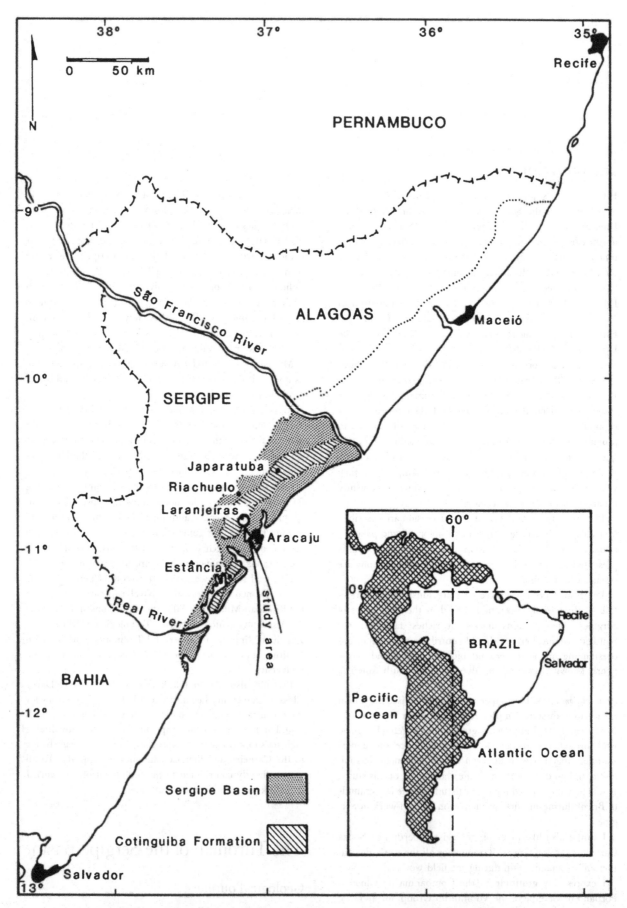

Fig. 1. Location of the Sergipe Basin, with Cotinguiba Formation indicated (continental Cenozoic Barreiras Group cover removed). (Modified from Bandeira 1978). Arrow points to area of study.

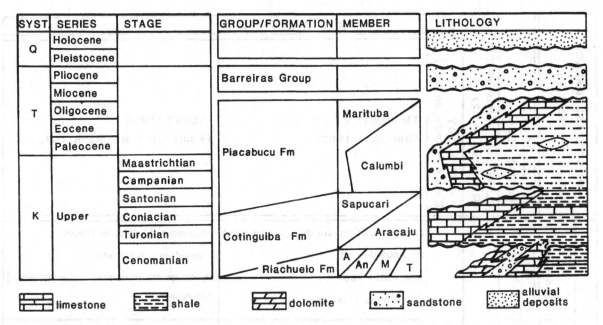

Fig. 2. Upper Cretaceous–Holocene stratigraphical framework of the Sergipe Basin (modified from Bengtson 1983 and Lana 1985). An – Angico Member; A – Aguilhada Member; M – Maruim Member; T – Taquari Member.

states of Sergipe and Alagoas (Fig. 1), between 9° and 11°30′S latitude and between 37°30′ and 35°30′W Greenwich longitude, comprising an area of approximately 16 000 km² of which some 10 000 km² is submerged. The sedimentary fill of the basin is up to 10 km thick and rests unconformably on the Precambrian basement complex (Ponte *et al.* 1980). With respect to its tectonic setting, the Sergipe Basin is of the Atlantic type (Asmus 1981), with a regional dip of 10° to 15° SE (Ojeda & Fugita 1976).

The Sergipe Basin is the southern half of the Sergipe–Alagoas Basin. The Mesozoic succession of Sergipe is one of the most complete among the marginal basins of the South Atlantic. The sequence starts in the Upper Jurassic with non-marine sediments and continues through the lowermost Cretaceous. Subsequently, in the Aptian, there was deposition from two evaporitic cycles corresponding to the first marine incursion (Asmus & Porto 1980; Ojeda 1983). This was followed by marine, clastic, shallow water sedimentation. There is no record of magmatic activity in the basin during the Mesozoic, such as that which can be found in the Santos, Campos (Asmus & Porto 1980), Pernambuco–Paraíba (Asmus 1982), and Amapá (Asmus & Guazelli 1981) basins.

The middle and upper Cretaceous are represented in the Sergipe Basin by three marine formations: the Riachuelo, Cotinguiba and Piaçabuçu formations (Fig. 2). The uppermost part of the Riachuelo Formation is locally Cenomanian, comprising mainly oolitic, oncolitic and pisolitic limestones and dolomites (Bandeira 1978). The Cotinguiba Formation was deposited in a lower energy environment with a low influx of terrigenous debris. It consists of shallow water carbonates (Sapucari Member), normally around 200 m thick, and of fine clastics, shales and limestones (Aracaju Member) which record the deepest deposition (ca 1 000 m) on the outer shelf and the continental slope (Freitas 1984). The Piaçabuçu Formation comprises shales and mudstones

(Calumbi Member), as well as pyrite-rich limestones and dolomites, with intercalations of sandy limestones (Marituba Member). This formation was deposited in a delta system and a carbonate platform. Both members range from the Campanian to the lower Pliocene (Ojeda 1984).

The Turonian is represented by a part of the Cotinguiba Formation which crops out in a 2–12 km wide band from the town of Japaratuba to the mouth of the Real River. The Sapucari Member, which has yielded the material studied here, is composed of massive and stratified limestones, locally with chert lenses and irregular nodules, and beds of breccia and coquina. The limestones are locally fairly dolomitic but rarely arenaceous. Within the Sapucari Member it is possible to recognize two depositional facies: Pindoba facies, with flaggy limestones, and Laranjeiras facies, with massive limestones (Bengtson 1983).

Biostratigraphy

Detailed biostratigraphical study of the Sergipe Basin started thirty years ago by the Companhia Brasileira de Petróleo S.A. (Petrobrás). The first attempt to establish a stratigraphic zonation of the Cotinguiba Formation based on macrofossils was by Petri (1962), who, based on K. Beurlen's (1961) work, subdivided the Turonian of Sergipe into three informal zones: *Vascoceras*, *Inoceramus labiatus* and *Sergipia*. G. Beurlen (1970) distinguished the 'Local Amplitude Zone' of *Coilopoceras* aff. *colleti* (lower Turonian).

Subsequently, Reyment & Tait (1972) and Reyment *et al.* (1976) presented a biostratigraphic zonation of the Cotinguiba Formation, where three ammonite zones could be recognized in the Turonian. Recently, Bengtson (1983) presented an exhaustive work on the Cenomanian–Coniacian sequence of Sergipe, in which he recognized three Turonian assemblages based on ammonite genera.

Meanwhile, micropalaeontologists have individually de-

	GROUPS OF FOSSILS				
AMMONITE ASSEMBLAGES (Bengtson 1983, 1986)	NANNOFOSSILS (Freitas 1984)	PLANKTONIC FORAMINIFERS (G.Beurlen 1982)	PALYNOLOGY (Lima & Boltenhagen 1981)	AMMONITES (Bengtson 1983)	INOCERAMIDS (Kauffman & Bengtson 1985)
TURONIAN — 3 (Subprionocyclus)	Lithastrinus grilli			Paralenticeras leonhardianum (Karsten) Parapuzosia sp. Reesidites spp. Subprionocyclus spp.	Inoceramus (I.) apicalis Woods I. perplexus Whitfield Mytiloides striatoconcentricus (Gümbel) Sergipia spp.
2 (Hoplitoides)		Hedbergella delrioensis	Gnetaceaepollenites crassipoli	Benueites? sp. Coilopoceras spp. Fagesia bomba Eck F. involuta Barber Hoplitoides ingens (von Koenen) H. gibbosulus (von Koenen) Kamerunoceras seitzi (Riedel) K. turoniense (d'Orbigny) Mammites nodosoides (Schlüter) Mitonia reesidei (Maury) Neoptychites cephalotus (Courtiller) Pachydesmoceras sp. Romaniceras deverianum (d'Orbigny) Spathites (S.) sp. Watinoceras jaekeli (Solger) W. amudariense (Arkhangel'skij) W. coloradoense (Henderson) W. guentherti Reyment	Inoceramus (I.) cuvieri Sowerby Mytiloides labiatus (von Schlotheim) M. mytiloides (Mantell) M. aff. hercynicus (Petrascheck) M. opalensis (Böse) M. subhercynicus (Seitz) Sergipia spp. "Sphenoceramus" spp.
1 (Pseudotissotia)				Mitonia spp. Nannovascoceras hartti (Hyatt) Pseudaspidoceras footeanum (Stoliczka) Pseudotissotia gabonensis Lombard P. nigeriensis plana Barber Thomasites sp. Vascoceras? globosum (Reyment) Wrightoceras sp.	Mytiloides mytiloides Mantell) M. ex. gr. submytiloides (Seitz)

Fig. 3. Correlation of the various systems for biostratigraphic subdivision or zonation of the Turonian of the Sergipe Basin (the most recent scheme for each fossil group is given). Bengtson's (1983, 1986) ammonite subdivision (left) corresponds to: 1 – the base of the Turonian; 2 – the major part of the Turonian; 3 – the uppermost part of the Turonian.

veloped Turonian biostratigraphical systems (Fig. 3), based on the study of nannofossils (Troelsen & Quadros 1971; Freitas 1984), planktonic foraminifers (Aurich *et al.* 1972; Noguti & Santos 1973; Sampaio & Northfleet 1975; G. Beurlen 1982) and palynomorphs (Regali *et al.* 1974; Herngreen 1975; Lima & Boltenhagen 1981).

Previous work on Turonian inoceramids from Brazil

This review summarizes original contributions in the fields of taxonomy and biostratigraphy.

The first account is due to Charles Frederick Hartt and dates back to 1868 (*A naturalist in Brazil*). In this work Hartt mentioned (p. 4) the presence of *Inoceramus* on the shores of Sergipe. In *Geology and Physical Geography of Brazil* Hartt (1870, pp. 383–384) quoted the existence of 'a great number of valves of a pretty *Inoceramus*, most probably new', in the limestones of the Sapucahy (=Sapucari) quarry, on the right-hand bank of the Cotinguiba River. In Chapter 19 (pp. 555–556), summarizing the geology of Brazil, Hartt tentatively placed the *Inoceramus* found in the laminated limestones (Cotinguiba Group) in the Senonian (associated with ammonites and fish remains).

Half a century elapsed before a new publication on this group of bivalves appeared. In 1925, Carlotta Joaquina Maury began her study of the Brazilian inoceramids with the publication of *Fosseis terciarios do Brasil, com descripção de novas formas cretaceas*. Out of the eleven species that have so far been described from Brazil, Maury described five species (and two subspecies).

From the Turonian only two species were reported up to 1980. These were found in the states of Pernambuco, Rio Grande do Norte and Sergipe and referred to *Inoceramus labiatus* von Schlotheim, 1813 and *I. (Sergipia) posidonomyaformis* Maury, 1925. The former was described by Maury (1937, pp. 110–113, Pl. 8:14) based on specimens from Morais Rego's collection (1922–1924?). The specimens are preserved in dense, massive, bluish and brown limestone from the village of Cedro on the margins of the Sergipe River (in the vicinity of localities Bumburum 4, 5 and 6 of Bengtson 1983). In addition, Maury described two new subspecies from the same locality: *I. labiatus cedroensis* Maury, 1937 and *I. labiatus sergipensis* Maury, 1937 (pp. 112–115, Pl. 8:12, 17). Duarte (1938, p. 34) reported several inoceramids from the lower Turonian of the Cotinguiba Formation, among them *I. labiatus* from localities near the bridge connecting the town of Laranjeiras and the village of Quitalé (probably between localities Laranjeiras 12 and 13 of Bengtson 1983). In 1970, Schaller (p. 72) mentioned the occurrence of the same species in the 'Local Amplitude Zone' of *Coilopoceras* aff. *colleti* (lower Turonian).

Inoceramus (Sergipia) posidonomyaformis was described and illustrated by Maury in 1925 (pp. 596–599, Pl. 22:6) based on specimens collected by Hartt from Sapucari. In 1937, she described and illustrated it in an identical manner (pp. 118–119, Pl. 8:15). On this occasion, Maury assigned the beds with *I. (Sergipia) posidonomyaformis* to the Maastrichtian, but both Morais Rego (1933, p. 56) and later Bender (1959, p.

17) dated them as Turonian. Duarte (1938, p. 45) reported on 'uma profusão de fósseis' ('a profusion of fossils') of this species, derived from the laminar limestones of the Cotinguiba Group, downstream from Cedro.

Inoceramus labiatus is also reported from Pernambuco and Rio Grande do Norte. The first author to observe *I. labiatus* in these states was K. Beurlen (1961) who found it to occur sparsely in the Beberibe and Sebastianópolis formations (the lowermost part of the present Jandaíra Limestone), both assigned at that time to the lower Turonian. Today, the Beberibe Formation is considered to be Santonian (K. Beurlen 1967a, 1967b). The specimens studied by K. Beurlen (1961) were collected in the Pernambuco–Paraíba Basin, north of Recife (pp. 43–44), and in the Potiguar Basin, between the villages of Açu and Upanema (pp. 45–46, 49). In his publication on the Jandaíra Limestone of the Mossoró region, K. Beurlen (1964) also referred to the collection of three specimens of *I. labiatus* from Mutamba do Arisco and Buraco d'Agua villages in the valley of the Açu River; brief taxonomic considerations and schematic drawings (Pl. 1:2) were also included.

The genus *Inoceramus* has been cited from Sergipe in a number of more general papers, such as Duarte (1936) who reported the presence of this genus in the Calumbi sandstone, 'directamente superposto ao calcareo lamellar, com o qual parece concordante' ('directly overlying the bedded limestones [Cotinguiba Formation], with which it appears concordant'). Duarte (1936. p. 117) assigned the Calumbi sandstone to 'the Danian stage, or possibly Eocene'. This was refuted by P.E. de Oliveira (1940, p. 3), who favoured a Cretaceous age. Kauffman (1977b) mentioned new specimens of *Sergipia posidonomyaformis* collected by Bengtson from 'Maury's localities'. Bengtson (1979) reported the presence of many inoceramid specimens that may provide an important item of study towards a better biostratigraphic understanding of the Brazilian Cretaceous basins. Bengtson (1983) and Kauffman & Bengtson (1985) listed these inoceramids and discussed the biostratigraphy. Bengtson & Berthou (1983), in a paper on microfossils of the Riachuelo and Cotinguiba formations, mentioned in Table 1 the occurrence of *Mytiloides mytiloides* and *M. submytiloides* in the limestones of the lower Turonian from Retiro 10, Sergipe.

Recently, Hessel (1986) described two new species from the lower Turonian of Retiro 15 (now included in Retiro 26), *Sphenoceramus mauryae* and *S. alatus*, which are redescribed herein.

Material and methods

One of the aims of this work was to investigate the biostratigraphic succession of inoceramids in the lower Turonian exposed in the largest limestone quarry of the Sergipe Basin. The quarry, which belongs to Companhia de Cimento Portland de Sergipe (CCPS), is here renamed Retiro 26.

During the first field-work, in January and February 1977, the exposures of the Cotinguiba Formation and the fossil assemblages in general were studied. The locality Retiro 26 was selected as the most suitable for a detailed study of the inoceramids. During subsequent visits to the area (June 1981 and March 1982) a systematic bed-by-bed collection

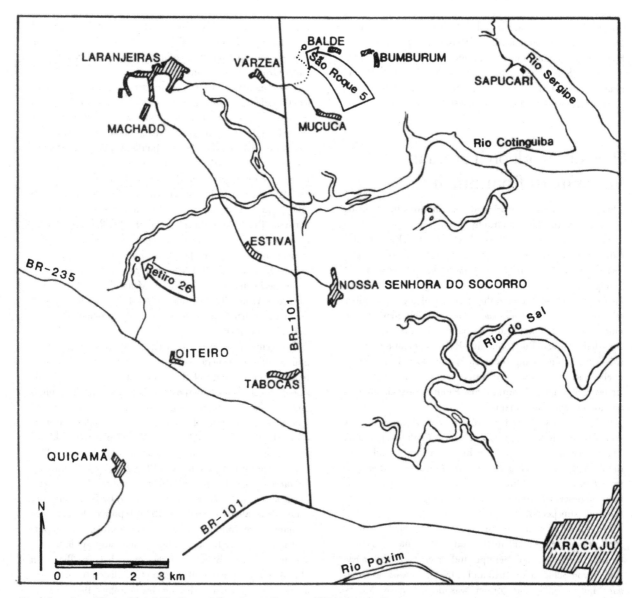

Fig. 4. Location of Retiro 26 and São Roque 5 (based upon Bengtson 1983, Appendix 3).

was made, with careful record of the lithology, stratigraphic boundaries, faunal assemblages and other geological data (such as thicknesses, dip of strata, etc). Upon completing the study of the fossils collected, it became clear that supplementary field-work was necessary for a more detailed sampling of divergently ornamented specimens and for the study of the occurrence of inoceramids in the adjacent area. This work was undertaken in February 1983.

The number of specimens collected from Retiro 26 amounts to approximately 150, to which were added eight other specimens collected from the same locality (then Retiro 15) by Suzana and Peter Bengtson (1971–1972), and kept at the Palaeontological Museum of Uppsala University. In general, the specimens are fairly well preserved, either as composite moulds ('Skultursteinkerne' of Seitz 1959) or as internal moulds with or without shell fragments. There are also shells whose inside shows the external ornamentation. In a few specimens, the articulation and the muscle scars can be seen. The specimens are usually damaged at the margins. The preservation of specimens with articulated

valves is rare. No preferred orientation in the deposition of the inoceramids could be observed. Weathering has not significantly affected the preservation. The specimens were prepared using a conventional vibrating drill.

The unpublished 1:25 000 topographic and geological maps of Petrobrás (see list in Bengtson 1983, p. 25) were used for the initial field-work. For the subsequent work the cartographic basis was a 1:25 000 topographic map prepared by COPEMI (Companhia Pernambucana de Mineração) for the CCPS.

In order to compare the Sergipe inoceramids with other forms from the Brazilian Turonian, I have studied the inoceramid collections at the Departamento Nacional da Produção Mineral, Rio de Janeiro, and at the Universidade Federal de Pernambuco, Recife – the only two Brazilian institutions possessing inoceramid specimens.

Since there is as yet no international agreement on the subdivisions of the Turonian Stage, these are treated here as informal units, as indicated by the lower-case letter of the modifyer (e.g., lower Turonian).

Fig. 5. View of the quarry Retiro 26, showing the smooth topography of the region (March 1982).

Locality descriptions

The quarry Retiro 26 lies 15 km northwest of Aracaju, the capital of the State of Sergipe, Brazil (Fig. 4). It includes the localities Retiro 15 and Retiro 16 of Bengtson (1983); it is also a synonym of Retiro 15 of Hessel (1986). The limestones of the Cotinguiba Formation have been exposed as the result of the quarrying activity of the CCPS. The quarry is oriented along a ENE–WSW axis, being well over 500 m in length and up to 200 m in width. The highest wall is up to 20 m high on the right-hand side of the quarry entrance. The topography consists in general of smoothly undulating hills with a maximum elevation of about 40 m (Figs. 5–7).

The localities are described using the system introduced by Bengtson (1983, pp. 30–31, 63):

RETIRO 26: UTM 8 800 400N/699 200E; UTM 8 800 800N/699 700E; UTM 8 800 250N/699 400E; UTM 8 800 700N/699 800E.
   Topographic sheet: SC.24-Z-B-IV Aracaju. Geological sheet: SC.24-Z-B-IV-4 Aracaju.
   Quarry oriented ENE–WSW. Altitude ca. 15–35 m.
   Kcsp: Hard, grey to light cream Laranjeiras limestones, massive to stratified, with breccia bed and several discontinuity surfaces.

The area adjacent to the quarry (within the limits of mining concession No. 50 043 of 24 January 1961 granted to CCPS), was investigated as far as the Retiro farm, covering approximately 2 250 × 1 250 m. Nineteen exposures yielded inoceramids, in beds consistently dipping southeast (Fig. 6). Fifteen of these localities were described by Bengtson (1983, pp. 69–70) under the denominations Retiro 1–9, 17–18 and 21, and Ribeira 15–17. Four new localities were studied and are described as follows.

RETIRO 22: UTM 8 800 800N/699 500E. Topographic sheet: SC.24-Z-B-IV Aracaju. Geological sheet: SC.24-Z-B-IV-4 Aracaju.
   Section on slope facing NW. Altitude ca. 15 m.

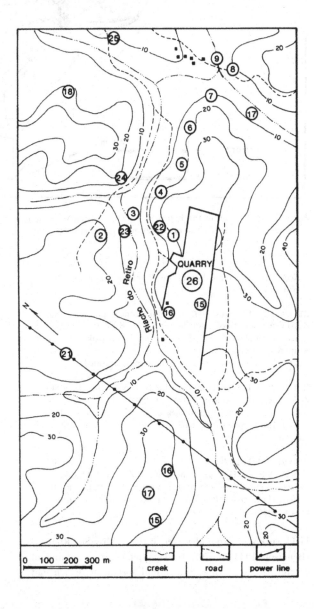

Fig. 6. Topographical map of the investigated area and location of the inoceramid localities. The exposures east of the power line are designated Retiro and those west of the line Ribeira (reduced to 1:12 500 from COPEMI's 1:25 000 topographical map).

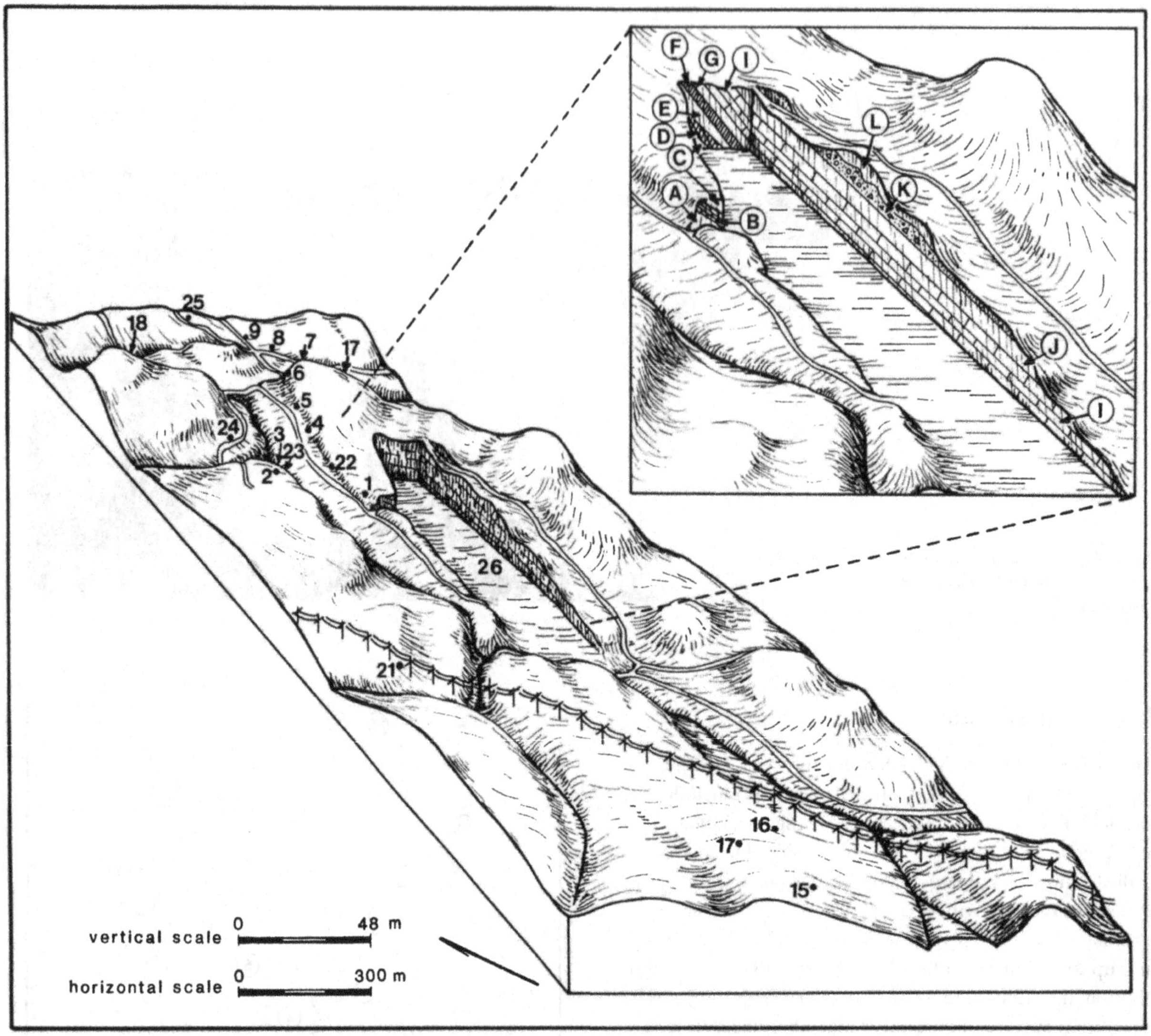

Fig. 7. Block-diagram of the Retiro 26 region, showing the limestone quarry with exposed lithological units, and the location of adjacent inoceramid localities. Vertical exaggeration of the inset diagram is 6.25.

Kcsp: Grey or cream Laranjeiras limestones, stratified, and of varying hardness.

RETIRO 23: UTM 8 800 900N/699 400E. Topographic sheet: SC.24-Z-B-IV Aracaju. Geological sheet: SC.24-Z-B-IV-4 Aracaju.
   Section on slope facing SE, on the left-hand bank of the creek. Altitude ca. 5 m.
   Kcsp: Hard, cream Laranjeiras limestones, stratified.

RETIRO 24: UTM 8 801 100N/699 600E. Topographic sheet: SC.24-Z-B-IV Aracaju. Geological sheet: SC.24-Z-B-IV-4 Aracaju.
   Section on hillslope and roadcut facing SE. Altitude ca. 10 m.
   Kcsp: Hard, cream Pindoba(?) limestones, stratified, with clayey intercalations.

RETIRO 25: UTM 8 801 850N/699 800E. Topographic sheet: SC.24-Z-B-IV Aracaju. Geological sheet: SC.24-Z-B-IV-4 Aracaju.
   Section on hillslope and roadcut facing SW. Altitude ca. 10 m.
   Kcsp: Hard, cream Pindoba(?) limestones, stratified, with clayey intercalations.

Field-work was also undertaken at the locality São Roque 5 (Bengtson 1983, p. 70), for complementary sampling of divergently ornamented inoceramid specimens, which had

been previously collected by P. and S. Bengtson (referred to '*Sphenoceramus*' by Kauffman & Bengtson 1985). The position of this locality is shown in Fig. 4.

Lithostratigraphy

The lithological subdivision of the Retiro 26 sequence was established in this work exclusively by depositional breaks. The strata have a constant dip to the southeast. The originally grey limestone develops a light cream colour upon weathering.

   The lithological units are as follows:

Bed A. – More than 4 m thick, it consists of alternating light and dark grey layers of massive hard limestone (Fig. 8); the layers become gradually brecciated towards the top of the unit (Figs. 8 and 9). Near the subrectilinear contact with the overlying bed B it returns to massive (Figs. 9 and 10). The clasts are subrounded and microcoquinoid, and their diame-

Fig. 8. Bed A, becoming gradually brecciated towards the top.

Fig. 9. Top of bed A (brecciated), bed B (massive) and bed C (breccia).

ter ranges from a few millimetres to approximately 40 cm. Dip varies from 9 to 15° SE. Crystalline calcite is commonly observed in irregular fillings and vugs.

Bed B. – Approximately 2 m thick, it consists of hard, massively bedded brownish-grey limestone (Fig. 9); it shows an irregular contact with the overlying bed C (Fig. 11).

Bed C. – With thicknesses varying from 1.5 to 4 m, it consists of conglomeratic to brecciated, massive, grey limestone; most clasts are decimetric but millimetric or metre-large clasts may also be present (Figs. 11–16). Figs 12 and 13 show a large, bioherm-like body with numerous irregular and branched perforations of *Thalassinoides* type. Locally, near the top, irregular black chert nodules with white rims are developed (Fig. 15). The contact with bed D is irregular and marked by an intraformational breccia (Fig. 17).

Bed D: Well bedded, greenish grey limestone, with thicknesses between 2.5 and 4 m; layers ranging from 10 to 20 cm thick, some more lithified than others (Fig. 16); dip varies from 9 to 15°SE.

Bed E: With approximately 3 m thickness, it consists of stratified, grey limestone with greenish partings; the base and top are microcoquinoid; bed E rests on an erosional surface with biogenic structures (Fig. 18). The bed is wedge-shaped, owing to a change of dip of the overlying unit (Fig. 19).

Fig. 10. Contact between beds A and B (detail of Fig. 9).

Bed F: Massive dark-grey and light-banded limestone, ca. 2 m thick, with subrectilinear lower contact (Fig. 20) and irregular upper contact marked by bleached halos around subrounded fragments (Fig. 21). Bed F dips 12 to 15° SE, making an angle of 3° with the underlying bed E (Fig. 19).

Fig. 11. Irregular contact between beds B and C.

Fig. 12. Bed C showing a large, bioherm-like body (middle right) disturbing the bedding of the overlying sediments.

Fig. 13. Bed C showing irregular and branched perfurations of biological origin on the bioherm-like body illustrated in Fig. 12.

Fig. 14. Bed C containing slightly rounded clasts of varying sizes.

Fig. 15. Bed C containing black chert nodules with white rims and small subrounded clasts.

Fig. 17. Contact between beds C and D, showing intraformational breccia (this is the contact illustrated by Bengtson 1983, Fig. 37).

Bed G: Ca 4 m thick, massive, dark-grey limestone with lighter intercalations (Fig. 19); dip 12–15° SE; the contact with the overlying bed is irregular, with biogenic structures appearing on the erosion surface (Fig. 22).

Bed H: Only 20 cm thick and formed by microcoquinoid limestone; its upper contact is subrectilinear, with clayey laminae (10 mm) (Fig. 22).

Bed I: This is the thickest (7.5 m) of all the beds at Retiro 26; it comprises dark and light layers of massive grey limestone (Fig. 23); a breccia level (Fig. 24) marks its upper limit with bed J.

Fig. 16. Irregular contact between beds C and D, influenced by large clasts. Note the distinct stratification of bed D.

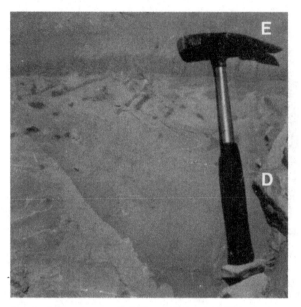

Fig. 18. Contact between beds D and E containing biogenic structures.

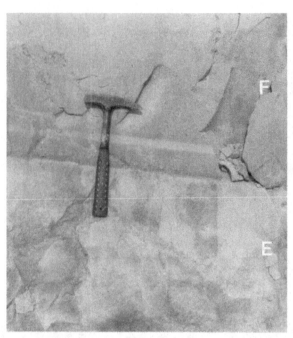

Fig. 20. Subrectilinear contact between beds E and F.

Bed J: Slightly over 2 m thick, light-cream, massive and hard limestone (colour due to weathering); the contact with bed K is irregular and marked by tubular ichnofossils on the erosion surface (Fig. 25).

Bed K: From 2 to 4 m thick, it consists of cream coloured limestone, locally with biogenic structures and irregular chert nodules similar to the bioherm-like body of bed C; the lithology is very friable and shows saccharoidal texture (Fig. 26); the contact with bed L is somewhat irregular.

Bed L: Ca. 2 m thick, deeply weathered, stratified cream coloured limestone with clayey intercalations (Fig. 27). This unit appears only in the uppermost part of the quarry, and is gradually transformed into soil (Fig. 26).

Taxonomic descriptions

The taxonomic descriptions employ the abbreviations of Matsumoto & Noda (1975, p. 200; 1985) for the morphology

Fig. 19. The wedge-shape of bed E, caused by the changing dip of the overlying F and G beds.

Fig. 21. Contact between beds F and G containing subrounded fragments with bleached halos.

of inoceramid shells, and for the convexity letter 'b' (Fig. 28). All measurements are given in millimetres. The specimens often have their margins damaged, which makes measurement difficult. Most of the specimens are moulds, many of which are deformed or flattened.

In the morphological descriptions, the ornament terminology is as follows (corresponding terms of Heinz 1928b in brackets): *cristae* ('Anwachskämme'), as part of a symmetric or asymmetric concentric undulation with sharp saddles; *rings* ('Anwachsringe'), as part of a concentric undulation with sharp lobes; *rugae* ('Anwachsrunzeln'), as part of a wholly smooth concentric undulation; *divergent ribs* ('Radialrippen'), as part of a straight or slightly curved divergent ornament with undulating cross-section; and *divergent rugae*,

Fig. 23. Bed I showing rhythmic banding of light and dark layers.

Fig. 22. Bed H showing an irregular contact with bed G. Biogenic structures can be seen on the lowermost contact. The contact with bed I is subrectilinear and marked by clayey laminae.

Fig. 24. Upper surface of bed I consisting of a breccia bed, which underlies bed J.

Fig. 25. Contact between beds J and K, with tubular ichnofossils.

Fig. 26. Bed K consisting of deeply weathered saccharoidal limestone.

as part of a crenulate divergent ornament with undulating cross-section (Fig. 29).

The holotypes are deposited in the collection of invertebrates of the Seção de Paleontologia of the Departamento Nacional da Produção Mineral (DNPM), Rio de Janeiro, under the numbers 6105–6108, 6055 and 6067. The paratypes and plaster casts of the holotypes are kept at the Palaeontological Museum of Uppsala University (PMU), under the numbers SA-149–227 and AF-404.

Class Bivalvia Linnaeus, 1758 pars.

Subclass Pteriomorpha K. Beurlen, 1944

Order Pterioida Newell, 1965

Superfamily Pteriacea Gray, 1847

Family Inoceramidae Giebel, 1852

Genus *Mytiloides* Brongniart, 1822 emend. Kauffman & Powell, 1977

Type species. – *Mytiloides labiatus* (von Schlotheim, 1813).
Diagnosis. – For a diagnosis and references see Kauffman & Powell (1977, pp. 71–72). In short, this genus is characterized by its medium size, thin shell, subequivalve and slightly convex shape, and its ovate and elongate outline.

Mytiloides mytiloides (Mantell, 1822)
Fig. 30A–B

Synonymy. – □ 1822 *Inoceramus mytiloides* – Mantell, p. 215, Pls. 27:3, 28:2. □ 1822 *Mytiloides labiatus*, A. Br. – Brongniart, pp. 81, 84, Pl. 3:4. □ 1911 *Inoceramus labiatus* (Schlotheim) 1813 – Woods, pp. 281–284, Pl. 50:2–3, 5 (not 50:1, 4, 6); Textfig.

Fig. 27. Bed L consisting of deeply weathered limestone with clayey horizons.

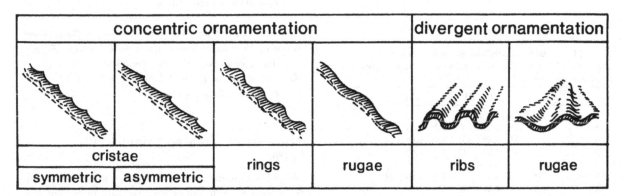

Fig. 28. Measurements taken on an inoceramid valve (according to Matsumoto & Noda 1975): S – hinge line; H – maximum dimension from the beak to the posteroventral extremity; L – maximum dimension of the line of measurement perpendicular to H; b – maximum valve convexity; l – shell length, i.e. the maximum dimension from anterior to posterior margins parallel to the hinge line; h – shell length, i.e. the maximum dimension perpendicular to l; d – angle between the hinge line and growth axis.

30. □ 1933 *Mytiloides labiatus* Schloth. – Heinz, pp. 248–249, Pl. 17:1–2 (not 17:3). □ 1935 *Inoceramus labiatus* var. *mytiloides* Mant. – Seitz, pp. 435–444; Figs. 2:a–d, 3:a–c, Pls. 36:1–4, 37:5 (not 37:4). □ 1962 *Inoceramus labiatus* Schlotheim – Hattin, p. 51, Pl. 14:G (not 14:B, D, F). □ 1977 *Mytiloides mytiloides* (Mantell) – Kauffman & Powell *in* Kauffman *et al.*, pp. 74–78, Pl. 6:12–15 (not 6:11, 16). □ 1978c *Mytiloides submytiloides* (Seitz) – Kauffman, pp. XIII.1–XIII.2, Pl. 1:2, 7–8. □ 1978c *Mytiloides mytiloides mytiloides* (Mantell) sensu Seitz 1934 – Kauffman, pp. XIII.1–XIII.2, Pl. 1:4, 12. □ 1978c *Mytiloides mytiloides* (Mantell) n. subsp. (late form, elongate shell) – Kauffman, pp. XIII.1–XII.2, Pl. 3:2. □ 1978c *Mytiloides labiatus* (Schlotheim) n. subsp. (elongate, finely ribbed, late form) – Kauffman, pp. XIII.1–XIII.2, Pl. 3:4 (not 3:5). □ 1978 *Mytiloides mytiloides* (Mantell) – Kauffman *in* Kauffmam *et al.*, p. XXIII.9, Pl. 10:8 (not 10:12). □ 1978 *Mytiloides mytiloides arcuata* (Seitz)? – Kauffman *in*

Kauffman *et al.*, p. XXIII.9, Pl. 10:9. □ 1978 *Mytiloides labiatus* (Schlotheim; sensu Seitz 1934) n. subsp. (late form) – Kauffman *in* Kauffman *et al.*, p. XXIII.9, Pl. 6:14. □ 1978 *I. labiatus mytiloides* Mantell – Robaszynski, pp. VIII.3–VIII.4, Pl. 2:4–6. □ 1982 *Mytiloides mytiloides* (Mantell, 1822) – Keller, pp. 121–125, Pl. 3:4, 6.

Material. – Six specimens preserved as composite moulds with shell fragments: three right and three left valves. All slightly fractured, chiefly at the ventral and posterior margins. PMU SA-199–202, 216 and 222.

Description. – The specimens are small to medium sized, thin shelled, and have moderate to low convexity and a rather inclined elongate umbo (Table 1). Auricle medium sized, subtriangular, flattened, and differentiated from the body of the shell by a marked umbonal fold, without, however,

concentric ornamentation				divergent ornamentation	
cristae		rings	rugae	ribs	rugae
symmetric	asymmetric				

Fig. 29. Schematic drawings of the different types of shell ornamentation in the inoceramids described in this paper (adapted from Heinz 1928b).

Fig. 30. Mytiloides spp. from Retiro 26, Sergipe: ×1. □ A-B. *M. mytiloides* (Mantell); A. (PMU SA-202); B. (PMU SA-222); □ C. *M. modeliaensis* (Sornay) (PMU SA-155); □ D. *M. submytiloides?* (Seitz) (PMU SA-195); □ E. *M.* aff. *mytiloides* (Mantell) (PMU SA-197).

forming a well defined sulcus (except for specimen PMU SA-199, but this could be due to post-mortem deformation). Umbo anterior, terminal, prosogyrous, slightly curved, projecting moderately above the hinge line. Anterior, ventral and posterior margins rounded, the latter more rectilinear. Juvenile and adult ornamentation distinct. Initially (10–17 mm from the umbo, measured along the growth axis) the ornament consists only of thin, subovate and subevenly spaced rugae, irregularly developed and without growth lines in between. Subsequently, the ornament consists of asymmetric cristae, more prominent, subequally developed and spaced (average 3.8–4.8 mm between adjacent cristae, as they spread out towards the posterior and ventral margin), with steeper distal flanks. Between the cristae, there are fine growth lines (three to five between adjacent cristae). Ontogenetic changes consist mainly of modifications in shell outline, from subovate to elongate-ovate, and changes in the characteristics of the umbo, from moderately to strongly prosogyrous, as well as the previously mentioned modifications in the shell ornament. Internal characters are not preserved. In specimen PMU SA-201, six minute ligamental pits are preserved in the damaged articulation.

Remarks. – The *Mytiloides mytiloides* specimens from Retiro 26 are similar to other specimens of this species known from elsewhere, in their elongate shell, rather inclined, moderately inflated, with a terminal umbo, triangular posterior auricle, and characteristic ornament consisting of regular concentric cristae with thin growth lines in the adult stage. On account of the thin shell and distinct juvenile stage (as well as the above mentioned features) the specimens are classified as belonging in *Mytiloides*. Furthermore, a large number of inoceramid fragments from Retiro 26 can probably be related to this species; they are not described here because of their poor state of preservation.

Occurrence. – Beds A and B at Retiro 26, Sergipe Basin; lower Turonian, in massive limestones of the Cotinguiba Formation (Laranjeiras facies). Field identifications were made, with some doubt, at Retiro 3, 5, 6, 7, 8, 9, 17, 18, 22, 23 and 24. The species is cosmopolitan.

Mytiloides aff. *mytiloides* (Mantell, 1822)
Fig. 30E

Material. – Two fragments of right valves preserved as composite moulds, with minute shell remains, umbo and dorsoanterior part preserved. PMU SA-196 and 197.

Table 1. Measurements of *Mytiloides mytiloides* from Retiro 26. RV – right valve; LV – left valve. Asterisk denotes inferred dimension of incomplete specimen.

Spec. No.	h	l	l/h	H	L	L/H	S	b	d
RV
SA-200	24*	31*	1.29	28*	26*	0.92	17*	3	47°
SA-201	36*	29*	0.80	43*	28*	0.65	23	2	43°
SA-222	48*	45*	0.93	51*	45*	0.88	24	5	51°
LV
SA-199	40*	46*	1.15	50*	33*	0.66	24*	3	47°
SA-202	80*	65*	0.81	67*	59*	0.88	–	6	-
SA-216	41*	45*	1.31	23*	43*	1.86	18	5	47°

Table 2. Measurements of *Mytiloides* aff. *mytiloides* from Retiro 26: RV – right valve. Asterisk denotes inferred dimension of incomplete specimen.

Spec. No.	h	l	l/h	H	L	L/H	S	b	d
RV									
SA-196	32*	32*	1.00	36*	32*	0.88	32?	12	37°
SA-197	38*	38*	1.00	48*	25*	0.52	15*	10	38°

Table 3. Measurements of *Mytiloides submytiloides*? from Retiro 26: RV – right valve. Asterisk denotes inferred dimension of incomplete specimen.

Spec. No.	h	l	l/h	H	L	L/H	S	b	d
RV									
SA-193	130*	124*	0.95	150*	56*	0.37	–	4	-
SA-195	60*	61*	1.01	39*	60*	1.53	–	5	-

Description. – The specimens are of moderate size, slightly inflated, obliquely ovate, and very elongate (Table 2). Auricle subtriangular, flattened, and separated from the rest of the shell by a marked umbonal fold. Umbo sharp, anterior, terminal, prosogyrous and projecting very slightly above the hinge line. Anterior margin slightly curved. Juvenile ornament poorly preserved. Adult ornament consists of faint, subregularly spaced (on average 5 mm), flattened and fairly uniformly developed, asymmetric cristae, with steeper distal flanks. Between the cristae there are generally 5 growth lines. Articulation and internal characters are not preserved.

Remarks. – The specimens resemble *M. mytiloides*, with their elongate, obliquely ovate shape, ornament with subregular cristae and five growth lines in between. They differ however in their very sharp and straight umbo, extremely faint cristae (showing a very delicate ornamentation), larger convexity and a well-defined umbonal fold separating the posterior auricle from the rest of the shell. They are similar to the *M. mytiloides* figured by Seitz (1935, Pl. 2c), but this specimen does not show the posterior auricle, which is well developed in the Sergipe specimens. The delicate ornament resembles that of *Inoceramus pictus* Sowerby, 1829, from the upper Cenomanian, although the Sergipe specimens have a sharper umbo and a more inclined shape. They are best considered as a form distinct from but closely related to *M. mytiloides*, although the formal establishment of a new species must await more and better preserved material.

Occurrence. – Bed B at Retiro 26, Sergipe Basin; lower Turonian, in massive limestones of the Cotinguiba Formation (Laranjeiras facies).

Mytiloides submytiloides? (Seitz, 1935)
Fig. 30D

Material. – Two fragments showing only part of the adult stage: a composite mould of a right valve, and a shell of another right valve. PMU SA-193 and 195.

Description. – Large, slightly convex, obliquely ovate valves (Table 3). In one of the specimens (PMU SA-195) a flattened, subtriangular posterior auricle can be seen, separated from the rest of the shell by a slightly prominent umbonal fold (no sulcus). The umbo, articulation, internal characters and juvenile ornament are not preserved. The adult ornament shows irregularly developed and spaced rugae (3–13 mm distance between adjacent rugae), in some cases bifurcating. These are generally not very prominent, slightly asymmetrical, wide and rounded. Growth lines are not normally visible; when present, however, they are thin, distant (average 1.5 mm), and number two to four between adjacent

rugae. Specimen PMU SA-193 shows a marked depression in the central part of the valve, which seems to be the result of physical injury of the organism while alive.

Remarks. – The specimens are too fragmentary to permit a definitive specific determination. Nevertheless, on account of the visible characters (only in the middle part of the organisms), they may be referred to *M. submytiloides* (Seitz, 1935) with a query, particularly on account of the obliquely ovate form and the irregular adult ornament.

Occurrence. – Beds A and B at Retiro 26, Sergipe Basin; lower Turonian, in massive limestones of the Cotinguiba Formation (Laranjeiras facies). Field identifications were made, with some doubt, at Retiro 2, 3, 6, 8 and 21. The species is cosmopolitan.

Mytiloides modeliaensis (Sornay, 1981)
Fig. 30C

Synonymy. – ☐ 1928 *Inoceramus labiatus* v. Schl. – Reeside, p. 1271, Pl. 10:7. ☐ 1981 *Inoceramus (Mytiloides) modeliaensis* n. sp. – Sornay, pp. 136–140, Pls. 1:1, 3–4, 2:1, 3–4.

Material. – Three right and one left valve preserved as internal moulds, with posterior and ventral margins broken, and part of the anterior margin and umbo fractured. The left valve has a counter-part covered with thin shell. PMU SA-154, 155, 194 and 213.

Description. – The specimens are small, subovate, somewhat obliquely elongate, not very convex (Table 4). The posterior auricle is a prolongation of the shell, and is separated from the body of the shell by a poorly marked umbonal fold. Umbo not very protruding, rounded, terminal, anterior and prosogyrous. Anterior margin rounded; ventral and posterior margins not preserved. The juvenile ornament consists of rounded asymmetrical cristae, regularly spaced, moderately salient. Density of cristae between seven and eight (up to 25 mm from the umbo). The adult stage is present only in one specimen, and the ornament consists of subregularly spaced,

Table 4. Measurements of *Mytiloides modeliaensis* from Retiro 26: RV – right valve; LV – left valve. Asterisk denotes inferred dimension of incomplete specimen.

Spec. No.	h	l	l/h	H	L	L/H	S	b	d
RV									
SA-154	26*	22*	0.84	25*	24*	0.96	14	6	50°
SA-194	82*	53*	0.64	77*	36*	0.46	11*	4	56°
SA-213	22*	25*	1.13	22*	24*	1.09	25	3	48°
LV									
SA-155	22*	27*	1.22	24*	23*	0.95	15	5	50°

Fig. 31. *Mytiloides* spp. from Retiro 26, Sergipe: ×1. □A-B. *M. transiens* (Seitz); A. (PMU SA-180); B. (PMU SA-178); □C. *M.* aff. *goppelnensis* (Badillet & Sornay) (PMU SA-164); □D-F *M. hercynicus* (Petrascheck); D. (SA-224); E. (PMU SA-177); F. (PMU SA-175).

ovate, flattened rings, with two to eight growth lines in between. The number of growth lines increases as the umbo recedes. The passage from the juvenile to adult stage is fairly well marked. In the adult stage the shell is more flattened and obliquely elongate-ovate. Articulation is not preserved.

Remarks. – The specimens of *Mytiloides modeliaensis* from Retiro 26 are similar to the initial part of the specimens described by Sornay (1981) from the lower Turonian of Quebrada La Modelia, Colombia, and similar to the one illustrated by the same author from eastern Ecuador (described by Reeside 1928). *Mytiloides duplicostatus* (Anderson, 1958) from the lower–middle Turonian of California is also similar, but this species is more acute and ovate, lacking well marked transition between the juvenile and adult stages. I examined the *Inoceramus labiatus sergipensis* (Maury 1937, pp. 112–115; Pl. 8:12), numbered 1163 and labelled 'holótipo', in the collections of the Departamento Nacional da Produção Mineral, Rio de Janeiro. It may be conspecific with or related to *M. modeliaensis*, taking into consideration the thin and obliquely ovate shell, the moderate convexity, the ornament with uniformly developed cristae, which are narrowly spaced and lack visible growth lines. However, in Maury's specimen the juvenile stage is poorly preserved, and the forms from Retiro 26 are almost entirely juveniles.

Occurrence. – Bed A of Retiro 26, Sergipe Basin; lower Turonian, in massive limestones of the Cotinguiba Formation

(Laranjeiras facies). Field identifications were made, with some doubt, at Retiro 1, 3, 7, 8, 9, 21 and 22. The species is known from Colombia, Ecuador, Mexico, Morocco; possibly also from Cedro (probably locality Bumburum 4 of Bengtson 1983), Sergipe Basin.

Mytiloides aff. *goppelnensis* (Badillet & Sornay, 1980)
Fig. 31C

Material. – Four right valves, preserved as composite moulds, with damaged posterior and ventral margins. PMU SA-164, 190, 192 and 198.

Description. – The specimens are obliquely elongate-ovate, moderately convex and medium sized (Table 5). Auricle of moderate size, flattened, subtriangular, with an undulation that is a prolongation of the adult ornament, and differentiated from the body of the shell by a shallow umbonal fold. Umbo not very sharp, terminal, prosogyrous, anterior, slightly projecting above the hinge line. Anterior margin rounded and almost vertical, and posterior margin subrectilinear. Ventral margin not preserved. Juvenile and adult ornament distinct. The former consists of subcircular rugae, subregularly developed and spaced, moderately salient, up to 12 mm away from the umbo (measured along the growth axis). There is a sudden change in the orientation of the ornament between the two ontogenetic stages, notably visible at the anterior margin, which bends considerably at this point. In the adult stage, the ornament becomes more salient, obliquely ovate, with nearly symmetric rings (the distal flank being slightly steeper), irregularly developed and spaced: at first (from 12–28 mm from the umbo) in a narrow pattern (on average 2 mm between adjacent rings), and finally in a wide pattern (on average 3 mm between adjacent rings). Growth lines are not visible or very faint (two or three lines) in between adjacent rings. Articulation and internal characters are not preserved. Ontogenetic changes

Table 5. Measurements of *Mytiloides* aff. *goppelnensis* from Retiro 26: RV – right valve. Asterisk denotes inferred dimension of incomplete specimen.

Spec. No.	h	l	l/h	H	L	L/H	S	b	d
RV									
SA-164	26*	30*	1.15	30*	30*	1.00	15*	08	50°
SA-190	33*	37*	1.12	40*	22*	0.55	–	02	–
SA-192	46*	50*	1.08	53*	50*	0.94	20*	10	52°
SA-198	36*	24*	0.66	31*	24*	0.77	10*	12	68°

are quite distinct, passing from a subequilateral, flattened juvenile shape to an adult stage, clearly inequilateral and more convex, along with the ornament modification from rugae to rings.

Remarks. – The specimens from Retiro 26 are similar to the specimen of *M. goppelnensis* figured by Badillet & Sornay (1980) and Sornay (1982) in their elongate shell and ornament pattern. They present a marked distinction between the juvenile and adult stages and a more irregular ornament (resembling the one found in *M. submytiloides*) than that of *M. goppelnensis*. *Mytiloides* aff. *goppelnensis* from Retiro 26 also has a much more salient, juvenile ornamentation which almost reaches the vertical anterior margin. They are best considered as a form distinct from but closely related to *M. goppelnensis*, although the formal establishment of a new species or subspecies must await more and better preserved material.

Occurrence. – Beds E and G at Retiro 26, Sergipe Basin; lower Turonian, in massive limestones of the Cotinguiba Formation (Laranjeiras facies).

Mytiloides hercynicus (Petrascheck, 1904)
Fig. 31D–F

Synonymy. – □ 1904 *Mytiloides hercynicus* n. sp. – Petrascheck, pp. 156–158; Textfig. 1, Pl. 8:2–3 (not 8:1). □ 1923 *Inoceramus Hercynicus* Petrascheck – Böse, pp. 181–183, Pl. 12:4–5 (not 12:1–3). □ 1928a *Inoceramus plicatus* D'Orb., var. *hercynica* Petr. – Heinz, pp. 65–68, Pl. 4:5. □ 1935 *Inoceramus labiatus* var. *hercynica* Petr. – Seitz, pp. 454–457. □ 1978c *Mytiloides hercynicus* (Petrascheck) – Kauffman, pp. XIII.1–XIII.2, Pl. 3:7. □ 1978 *Mytiloides subhercynicus* (Seitz) n. subsp. transitional to *M. mytiloides* (Mantell) – Kauffman *in* Kauffman *et al.*, p. XXIII.9, Pl. 6:12 (not 6:10). □ 1982 *Inoceramus hercynicus* Petrascheck – Sornay, Pl. 1:2 (not 1:1a, 1c). □ 1982 *Mytiloides hercynicus* (Petrascheck, 1903) – Keller, p. 131, Pl. 4:1.

Material. – Seventeen valves preserved as internal or composite moulds with shell fragments: eight left valves, eight right valves and one specimen with slightly dislocated valves (PMU SA-226). All slightly fractured, usually in the ventral, anterior or posterior margins. PMU SA-175–177, 179, 182, 184–186, 188, 189, 191, 223, 224, 226 and 230.

Description. – The specimens are obliquely subovate, with moderate to low convexity, and of small to medium size (Table 6). Posterior auricle (only preserved in some specimens) subtriangular, of moderate length, with prolongations of the adult ornament. Umbo anterior, terminal, not very prosogyrous, rounded, flattened, not projecting above the hinge line. Anterior and posterior margins rounded; ventral margin not preserved. Juvenile ornament occasionally indistinct from that of the adult stage, consisting of uniformely developed, subregularly and closely spaced cristae. In adult stage the concentric ornament results in more widely spaced rings (average 3.6 mm between adjacent rings), more salient, undulated, slightly asymmetric (distal flank slightly steeper), with thin growth lines (two to six between adjacent

Table 6. Measurements of *Mytiloides hercynicus* from Retiro 26: RV – right valve; LV – left valve. Asterisk denotes inferred dimension of incomplete specimen.

Spec. No.	h	l	l/h	H	L	L/H	S	b	d
Bivalved									
SA-226									
RV	27*	21*	0.77	29	17*	0.58	–	3	-
LV	22*	17*	0.77	26*	09*	0.34	–	2	-
Univalved									
RV									
SA-176	24*	18*	0.75	26*	14*	0.53	06*	3	48°
SA-177	40*	36*	0.90	32*	34	1.06	18	2	50°
SA-184	50*	60*	1.20	50*	41*	0.82	21*	2	48°
SA-188	58*	41*	0.70	52*	46*	0.88	13*	5	50°
SA-189	32*	35*	1.09	36*	31*	0.86	–	3	-
SA-224	38*	40*	1.05	49*	36*	0.73	16*	1	51°
SA-230	24*	35*	1.45	32*	26*	0.81	–	2	-
SA-233	08*	08*	1.00	08*	07*	0.87	04*	1	60°
LV									
SA-175	18*	16*	0.88	22*	13*	0.59	07	4	53°
SA-179	20*	21*	1.05	11*	19*	1.72	08	2	54°
SA-182	43*	34*	0.79	38*	30*	0.79	16*	1	59°
SA-185	43*	54*	1.25	45*	45*	1.00	–	5	-
SA-186	19*	18*	0.94	21*	12*	0.57	07	2	48°
SA-191	36*	38*	1.05	41*	34*	0.82	–	6	-
SA-216	42*	57*	1.35	70*	42*	0.60	25	5	49°
SA-223	23*	21*	0.91	24*	20*	0.83	–	4	-

rings). The specimen PMU SA-185 shows very faint adult rings. PMU SA-182, 184 and 224 have flattened valves, which is probably a post-mortem effect. Articulation and internal characters are not preserved. Ontogenetic changes consist mainly of modifications in shell outline and ornamentation, from subcircular with cristae to elongate-ovate with rings. The passage from the juvenile to adult stage is not very well defined, but the presence of rings and growth lines marks the adult stage.

Remarks. – The specimens from Retiro 26 listed here have morphologic features (obliquely ovate, moderately convex shell, with a rounded umbo and rings and growth lines in adult ornament) similar to *Mytiloides hercynicus* known from elsewhere in the world.

Occurrence. – Beds E, F, G and I at Retiro 26, Sergipe Basin; lower Turonian, in massive limestones of the Cotinguiba Formation (Laranjeiras facies). The species is cosmopolitan.

Mytiloides transiens (Seitz, 1935)
Fig. 31A–B

Synonymy. – □ 1923 *Inoceramus hercynicus* Petrascheck – Böse, pp. 181–183, Pl. 12:3 (not 12:1–2, 4–5). □ 1935 *Inoceramus labiatus* n. var. *subhercynica* – Seitz, pp. 465–469, Textfig. 18a–18f; Pl. 40:1–2. □ 1935 *Inoceramus labiatus* var. *subhercynica* n. forma *transiens* – Seitz, pp. 465–469, Pl. 40:3. □ 1978c *Mytiloides subhercynicus transiens* (Seitz) – Kauffman, pp. XIII.1–XIII.2, Pls. 1:6, 2:2, 7. □ 1982 *Mytiloides transiens* (Seitz, 1934) – Keller, p. 133, Pl. 3:5.

Material. – Five specimens preserved as composite moulds with scattered shell fragments: four right and one left valve. All specimens slightly fractured, chiefly on the posterior and

Table 7. Measurements of *Mytiloides transiens* from Retiro 26: RV – right valve; LV – left valve. Asterisk denotes inferred dimension of incomplete specimen.

Spec. No.	h	l	l/h	H	L	L/H	S	b	d
RV									
SA-152	24*	33*	1.37	20*	33*	1.65	21*	2	44°
SA-166	22*	28*	1.27	30*	27*	0.90	16	5	50°
SA-180	58*	58*	1.00	70*	48	0.68	24	6	48°
SA-181	34*	52*	1.52	45*	36*	0.80	23*	3	45°
LV									
SA-178	28*	43*	1.53	43*	31*	0.72	22	3	41°

Table 8. Measurements of *Sergipia* aff. *posidonomyaformis* from Retiro 26: RV – right valve; LV – left valve. Asterisk denotes inferred dimension of incomplete specimen.

Spec. No.	h	l	l/h	H	L	L/H	S	b	d
RV									
SA-183	15*	17*	1.13	16*	16*	1.00	05*	1	62°
SA-211	21*	25*	1.19	20*	20*	1.00	06*	1	60°
SA-212	21	20*	0.95	23*	19*	0.82	06	2	69°
SA-217	25*	21*	0.84	26*	19*	0.73	06*	1	56°
LV									
SA-170	10*	13*	1.30	11*	13*	1.18	07	1	61°
SA-210	21*	30*	1.42	24*	21*	0.87	10*	1	50°
SA-213	13*	30*	2.30	14*	22*	1.57	10*	1	55°
SA-214	15*	23*	1.53	20*	24*	1.20	10*	2	45°
SA-221	31*	33*	1.06	37*	32*	0.86	06	1	65°

ventral margins, rarely on the anterior. PMU SA-152, 166, 178, 180 and 181.

Description. – The specimens are subovate, obliquely elongate, medium-sized, and moderately convex (Table 7). Posterior auricle flattened, subtriangular, well developed, separated from the rest of the shell by a marked umbonal fold, which does not form a sulcus. Umbo not very sharp, terminal, prosogyrous, anterior, flattened, not projecting over the hinge line. Posterior, ventral and anterior margins rounded; the posterior becoming subrectilinear. Anterior margin almost vertical. Juvenile ornament distinct from adult ornament, consisting of thin subcircular rugae, subregularly spaced, starting up to 18–22 mm from the umbo (measured along the growth axis). In adult stage the ornament becomes more salient, with asymmetrical rings (the distal flank steeper), subregularly developed and spaced (on average 2 mm between adjacent rings); one to four growth lines between adjacent rings; the growth lines may be obliterated depending on the state of preservation. Internal characters and articulation are not preserved. Ontogenetic changes consist mainly of a modification in the growth axis and ornament, from the rounded outline (with rugae) of the juvenile stage to the elongate outline (with rings and growth lines) of the adult stage.

Remarks. – The specimens from Retiro 26 are similar to other *Mytiloides transiens* (Seitz, 1935) found elsewhere in the world. Considering the thin shell and other morphological features referred by Kauffman & Powell (1977) to the genus *Mytiloides*, I attribute this form to *Mytiloides transiens*. It is more elongate than *M. hercynicus*, with a more abrupt change in the shape between the juvenile and the adult stages, and with more closely spaced concentric ornamentation. A number of fragments collected at Retiro 26 may also belong to this species.

Occurrence. – Beds E to I at Retiro 26, Sergipe Basin; lower Turonian, in massive limestones of the Cotinguiba Formation (Laranjeiras facies). Cosmopolitan.

Genus *Sergipia* Maury, 1925

Type species. – *Sergipia posidonomyaformis* (Maury, 1925).

Diagnosis. – For the diagnosis and references, see Cox (1969, p. N320). The characteristics of this genus are the small to medium sized, thin, flattened, rounded shell with very delicate concentric ornamentation.

Sergipia aff. *posidonomyaformis* (Maury, 1925)
Fig. 32A–C

Material. – Nine specimens preserved as entire shells (PMU SA-183, 210, 212 and 217) or as composite moulds with shell fragments: four right and five left valves. All slightly fractured. PMU SA-170, 183, 210–214, 217 and 221.

Description. – The specimens are small, very thin-shelled, equivalve, very flattened, subovate, slightly prosocline (Table 8). Hinge line subrectilinear and forming obtuse angles with the anterior and posterior margins, which are both rounded. Small anterior auricle with a slight umbonal fold. Posterior auricle more developed, subtriangular, separated from the rest of the shell by a more accentuated umbonal fold. Umbo anterior, slightly prosogyrous, very flat, subcentral. The ornament consists of concentric, undulating, subcircular, regular and closely spaced rings on the entire surface of the shell. Between the rings there are growth lines (two to seven) beginning at a certain ontogenetic stage. At the end of the juvenile stage there are two to three growth lines between the rings. Only one specimen (PMU SA-170) lacks growth lines. The average distance between the concentric rings is 1.2 mm, up to 20 mm from umbo (measured along the growth axis). Ontogenetic changes consist mainly of a posterior elongation of the shell and a modification of the ornament, which becomes stronger and with a greater number of growth lines. The passage from the juvenile to adult stage is relatively well marked. Internal characters and articulation are not preserved.

Remarks. – The specimens show very close affinity with *Sergipia posidonomyaformis* (Maury, 1925), especially as regards the flattened, subovate shape, the thin shell, the subcentral umbo and the regular and closely spaced rings. But the Retiro 26 specimens have more distinct auricles, a subrectilinear hinge line and, at the end of the juvenile stage, growth lines between concentric rings. The holotype of *Sergipia posidonomyaformis* is very small (l = 20 mm and h = 22 mm: Maury 1925), and may therefore not have had time to develop growth lines.

Inoceramus (*Sergipia*) *posidonomyaformis* var. *scheibei* Heinz (1928a, pp. 84–85) from Mexico is an elongate form, recalling the general shape of *Mytiloides*, according to Heinz' illustrations (1928, Pls. 4:6, 5:6). There is only little similarity with the specimens from Retiro 26. Kauffman (1977b)

Fig. 32. Sergipia spp. from Retiro 26, Sergipe, ×1. □ A-C. *S.* aff. *posidonomyaformis* (Maury); A. (PMU SA-221); B. (PMU SA-217); C. (PMU SA-210); □ D-F. *S. hartti* n. sp.; E. holotype (DNPM 6107); E. (PMU SA-218); F. (PMU SA-208); □ G. *S.* sp. (PMU SA-203).

mentioned the similarity of Heinz' subspecies with *Mytiloides teraokai* Matsumoto & Noda, 1968 from the middle Turonian of southwest Japan. On the other hand, *I.* (*Sergipia*) *posidonomyaformis*, also described by Heinz (1928a, pp. 83–84, Pl. 5:3, 5) from Mexico, may be compared with the species described here, but in the former the umbo, auricles and shell material are not preserved.

The Retiro specimens also show some similarity to *Sergipia akamatsui* (Yehara) [*Inoceramus* (*Sergipia*?) *akamatsui* Yehara, 1924, Pl. 2:2, 4] from the middle Turonian to basal Coniacian of Japan. But this species is much smaller, with sporadic, divergent grooves, and lacks dorsal auricles (which may have been broken, as suggested by Kauffman 1977b).

Kauffman is currently revising the genus *Sergipia* as well as other species from the Cotinguiba Formation of Sergipe. This work will certainly contribute to a better knowledge of the possible intraspecific variability of *Sergipia posidonomyaformis*.

Occurrence. – Beds E to J at Retiro 26, Sergipe Basin; lower Turonian, in massive limestones of the Cotinguiba Formation (Laranjeiras facies).

Sergipia hartti n. sp.
Figs. 32D–F, 33, 34

Synonymy. – □v 1976 *Sergipia* n. sp. aff. *posidonomyaformis* Maury – Offodile, p. 83, Pl. 16:2. □v 1977 *Sergipia* aff. *posidonomyaformis* Maury – Offodile & Reyment, p. 47, Fig. 15.

Derivation of name. – In honour of the Canadian scientist Charles Frederick Hartt, who was the first to mention the presence of inoceramids in Sergipe (Hartt 1868).

Holotype. – Specimen DNPM 6107, collected by Maria Helena Ribeiro Hessel in March 1982.

Type locality. – Retiro 26, 15 km NW of Aracaju, Sergipe, Brazil.

Type stratum. – Lower Turonian, Cotinguiba Formation, bed

I of the section here studied. The level falls within the *Turonian 2* ammonite assemblage of Bengtson (1983).

Material. – Including the holotype, twelve specimens (one with both valves, two right and eight left valves) well preserved as composite moulds or as shell fragments (PMU SA-187, 215, 219 and 220RV), with little material loss at the margins. A left valve from Nigeria (PMU AF-404, see below) is also included among the paratypes. The holotype is deposited in the Seção de Paleontologia of the Departamento Nacional da Produção Mineral, Rio de Janeiro, under number 6107. A plaster cast is kept in the Palaeontological Museum of Uppsala University, under number SA-209. The paratypes are at the same institution (Uppsala) under numbers SA-151, 187, 204–208, 215, 218–220 and AF-404.

Diagnosis. – Medium-sized, equivalve, inequilaterally elongate, very flattened, thin-shelled. Prosogyrous anterior umbo. Very distinct, elongate posterior auricle. Regular and closely spaced concentric rings, between which thin growth lines appear. Hinge line of moderate length. Posterior, ventral and anterior margins rounded.

Description. – The holotype is a left valve preserved as a composite mould, with the umbo slightly damaged, and the

Fig. 33. Sergipia hartti n. sp. (PMU AF-404) from Nigercem Quarry, Nkalagu, Anambra State, Nigeria. ×1.5.

Table 9. Measurements of *Sergipia hartti* from Retiro 26: RV – right valve; LV – left valve. Asterisk denotes inferred dimension of incomplete specimen.

Spec. No.	h	l	l/h	H	L	L/H	S	b	d
Bivalved									
SA-220									
RV	38*	53*	1.39	52*	25*	0.48	–	1	-
LV	34*	52*	1.52	40*	37*	0.92	–	1	-
Univalved									
RV									
SA-205	17*	23*	1.35	19*	21*	1.10	04*	2	58°
SA-208	32*	44*	1.37	32*	33*	1.03	15	1	59°
LV									
SA-151	22*	22*	1.00	21*	19*	0.90	05*	1	52°
SA-187	23*	25*	1.08	26*	26*	1.00	07*	1	56°
SA-204	42*	46*	1.09	36*	45*	1.25	20	1	65°
SA-206	23*	36*	1.56	24*	34*	1.41	14*	2	60°
SA-207	42	43*	1.02	46	38*	0.82	12*	2	50°
SA-215	16*	26*	1.62	23*	22*	0.95	09	2	58°
SA-218	21*	25*	1.19	25*	21*	0.84	05*	1	55°
SA-219	38*	49*	1.28	40*	42*	0.95	12*	1	63°
DNPM-6110	24	32	1.33	31	28*	0.90	12	3	48°
(SA-209)									

posterior and ventral margins slightly broken. There are small fragments of thin shell material at the anterior margin. Measurements are given in Table 9. Umbo anterior, prosogyrous, slightly above the hinge line, which is rectilinear and of moderate length. Shell ornament consists of undulating, regular, closely spaced concentric rings becoming narrower near the umbo. Average distance of the rings 2 mm at a distance of 10–25 mm from the umbo. At 18 mm from the umbo (measured along the growth axis), two or three thin

growth lines become intercalated with each concentric ring. Small and elongate posterior auricle separated from the rest of the shell by a well marked rectilinear umbonal fold. Adult rings continue to the auricle. Ontogenetic changes consist mainly of ornament modifications and increase in shell elongation, although the shell is inequilateral throughout the ontogeny (Fig. 34). Articulation not preserved. There are small depressions from physical injuries in the central part of the valve.

Remarks on the paratypes. – The holotype is the best preserved specimen, with minor damage and well-preserved morphological details. The variation of the L/H ratio and the angles are illustrated in Fig. 34. There is significant variation in the size of the shell and the auricle, owing to the fragmentation of the posterior margin. The posterior auricle forms an obtuse angle with the cardinal margin. Specimens larger than the holotype are common (Table 9). In general, the growth lines appear at a distance of between 18 and 20 mm from the umbo. There may be up to four growth lines between adjacent rings. In specimen PMU SA-205 the growth lines are not visible. The ornament in the adult stage is more flattened and the growth lines are more salient than in the juvenile stage. There is very little distinction in the passage between the two ontogenetic stages. The anterior margin is abruptly terminated and the posterior region is flattened. Specimen PMU SA-207 shows a large, elongate muscle scar in the posterior region, probably pertaining to the adductor muscle. The specimen PMU SA-206 is juvenile form.

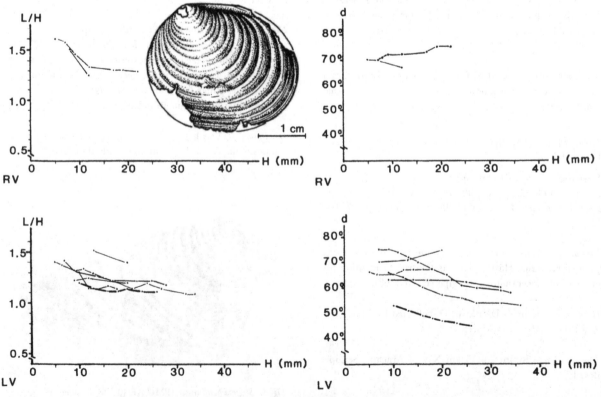

Fig. 34. Sergipia hartti n. sp. Ontogenetic variations of L/H ratio and angle d; figure shows holotype (DNPM 6107).

Affinities. – *Sergipia hartti* is a species which, in spite of its very thin flattened shell and delicate ornamentation, is markedly inequilateral, with a lateral umbo and only one elongate posterior auricle, differing in these features from other species of *Sergipia* known to date. The Nigerian specimen of *Sergipia* aff. *posidonomyaformis* illustrated by Offodile (1976) and Offodile & Reyment (1977) was examined by me (specimen AF-404 of the collection of the Palaeontological Museum of Uppsala University). It is a *S. hartti*, on account of its shape and ornament. The specimen is a mould of a left valve (Fig. 33) and has the following measurements (linear measurements in mm; * denotes inferred dimensions): h = 31*; l = 50*; l/h = 1.61; H = 30* (juvenile only); L = 28*; L/H = 0.93; S = 10*; b = 2; d = 48°. It has asymmetric rings with growth lines between them in adult stage. The juvenile stage shows fine divergent lines, a character that suggests, with the thin flattened and inequilateral shell, a phylogenetic relationship with the Coniacian genus *Didymotis*. This very well preserved specimen from Nigeria also shows part of the articulation, with twelve ligamental pits, which are more closely spaced near the umbo.

Occurrence. – Beds E to I at Retiro 26, Sergipe Basin; lower Turonian, in massive limestones of the Cotinguiba Formation (Laranjeiras facies), associated with the ammonites *Benueites*? sp. and *Watinoceras*? sp. The species also occurs in Nigeria (basal Ezeaku Formation).

Sergipia sp.
Fig. 32G

Material. – One right valve preserved as an internal, possibly composite, mould with the ventral margin slightly fractured. PMU SA-203.

Description. – Inequilateral, ovate, very flattened valve, with a rectilinear hinge line, which forms obtuse angles with the anterior (110°) and posterior (148°) margins. The valve dimensions are (linear measurements in mm; * denotes inferred dimensions): h = 24*; l = 33*; l/h = 1.37; H = 31*; L = 26; L/H = 0.84; S = 14; b = 2; d = 65°. Anterior auricle short, forming an angle of approximately 52° with the hinge line; a slight umbonal fold separates the latter from the rest of the shell. Posterior auricle not very pronounced, reduced. Prosogyrous, subterminal, flattened umbo, not projected above the hinge line. Shell ornament consists of semicircular, concentric, symmetric cristae, which are very narrowly (average distance 1 mm), subregularly spaced. Internal characters and articulation are not preserved. Ontogenetic changes consist of modifications of the shell outline, from semicircular to semiovate, and in the position of the umbo, from subcentral to sublateral.

Remarks. – *Sergipia* sp. differs from the other species of the genus by its subterminal umbo, very narrow-spaced, thin, concentric cristae, distinct anterior auricle and its reduced posterior auricle. It probably represents a new species; however, since only one, poorly preserved specimen is available, a new species is not formally established.

Occurrence. – Bed G at Retiro 26, Sergipe Basin; lower Turon-

ian, in massive limestones of the Cotinguiba Formation (Laranjeiras facies).

Genus *Rhyssomytiloides* n. gen.

Derivation of name. – From Greek *rhyssos*, ruga, for the conspicuous rugae arranged in a divergent pattern, and *Mytiloides*, a genus characterized by its thin and elongate shell. The name is masculine.

Type species. – *Rhyssomytiloides mauryae* (Hessel, 1986).

Diagnosis. – Small to moderate size, inequilateral, subequivalve, subrounded to elongate-ovate, with moderate to high convexity. Prosogyrous, terminal umbo. Hinge line moderate to long. Small posterior auricle. Ornamentation consisting of concentric cristae which at a certain growth stage become superposed with divergent ribs or rugae. Thin shell.

Comparison. – The type of ornamentation that characterizes *Rhyssomytiloides*, i.e., divergent ribs and rugae, also occurs in other Cretaceous inoceramids. *Rhyssomytiloides* differs from these in the following respects:

From *Sphenoceramus* Böhm, 1915, known from Santonian and Campanian strata of the North Temperate Realm (Kauffman 1973; Dhondt 1983), *Rhyssomytiloides* differs in being smaller and having elongate-ovate, obliquely inclined outline (not trigonal), umbo less sharp, thinner shell and more delicate concentric ornament.

Rhyssomytiloides also differs from *Birostrina* Sowerby, 1821 in being subequivalve, with a median hinge line, more protruding umbo, divergent ribs or rugae present also on the anterior and posterior regions of the shell. *Birostrina* is a cosmopolitan Albian–Cenomanian genus.

Spyridoceramus Cox, 1969, from the Maastrichtian of Europe and North America, has subregular, thin and closely spaced divergent undulations on the entire surface, an anterior auricle, and a markedly protruding umbo. These features are not seen in *Rhyssomytiloides*.

Rhyssomytiloides differs from *Retroceramus* (*Striatoceramus*) Koschelkina, 1960 (*fide* Cox 1969), known from the Middle Jurassic of Siberia, because it has a less elongate and oblique shell, less prominent umbo and more delicate, divergent ornament.

Rhyssomytiloides can also differ from *I.* (*Mytiloceramus*) Rollier, 1914, which occurs in the Middle Jurassic of Europe and Japan, because the former does not have an anterior auricle, having more flattened concentric ornament and more divergent ribs or rugae, which form a nodule at the point of intersection.

Rhyssomytiloides can also be compared with the cosmopolitan *Mytiloides* Brongniart, 1822 emend. Kauffman & Powell, 1977, on account of the general shell outline and convexity of the juvenile stage, the elongate shape in the adult stage and medium sized, thin shell. Nevertheless, *R.* has well developed divergent ribs or rugae, concentric ornamentation with similar juvenile and adult stages, and a less distinct auricle. The shell is occasionally considerably more convex than typical *Mytiloides*.

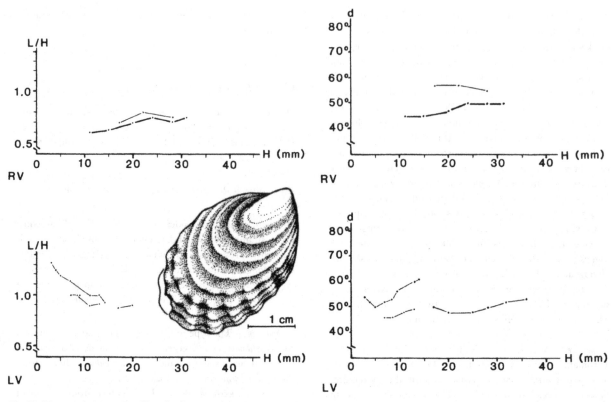

Fig. 35. Rhyssomytiloides mauryae (Hessel). Ontogenetic variations of L/H ratio and angle d; figure shows holotype (DNPM 6067).

Species of Rhyssomytiloides. – *R. mauryae* (Hessel, 1986), *R. bengtsoni, R. alatus* (Hessel, 1986), *R. beurleni* and *R. retirensis.*

Occurrence. – In Retiro 17, Retiro 21 and São Roque 5 (localities of Bengtson 1983), Sergipe Basin, Brazil; lower Turonian, in massive limestones of the Cotinguiba Formation (Laranjeiras facies), associated with *Mytiloides mytiloides*, *M. modeliaensis*, and the ammonites *Mammites* sp., *Neoptychites* sp. and *Kamerunoceras* sp. (ammonite assemblage Turonian 2 of Bengtson 1983). The genus also occurs in Madagascar and South West Africa.

Rhyssomytiloides mauryae (Hessel, 1986)
Figs. 35, 36A–C

Synonymy. – □v 1934 *Cladoceramus* cf. *diversus* Stoliczka – Basse, pp. 87, 93–95. □ 1977 *Sphenoceramus* aff. *S. schmidti* – Klinger, pp. 91, Fig. 8H. □v 1986 *Sphenoceramus mauryae* n. sp. – Hessel, pp. 228–234, Pl. 1:1a–b, 1:2, 1:3, Textfig. 5.

Derivation of name. – In honour of the North American pa-laeontologist Carlota Joaquina Maury, who described Brazilian inoceramids for the first time (1925).

Holotype. – Specimen DNPM 6067, collected by Albertino de Souza Carvalho in June 1981.

Type locality. – Retiro 26, 15 km NW of Aracaju, Sergipe, Brazil.

Type stratum. – Lower Turonian, Cotinguiba Formation, bed A of the section here studied. The level falls within the Turonian 2 ammonite assemblage of Bengtson (1983).

Material. – Including the holotype, three right and three left valves, well preserved as internal or composite moulds, with minute shell fragments close to the articulation and to the posterior margin. The specimen PMU SA-172 has a counterpart with shell. A bivalved specimen from Madagascar examined as a plaster cast and described below is also included among the paratypes. The holotype is deposited in the Seção de Paleontologia of the Departamento Nacional da Produção Mineral, DNPM (Rio de Janeiro) under number 6067. A plaster cast is kept in the Palaeontological Museum of Uppsala University, under number SA-174. The paratypes are at the same institution (Uppsala) under numbers SA-149, 171–173 and 225.

Diagnosis. – Small, thin-shelled, inequilateral, elongate-ovate, with moderate to high convexity. Regularly developed and spaced asymmetrical concentric cristae, superposed by strong regular divergent ribs, which appear rather late in the shell growth. Rounded and crenulated margins. Prosogyrous and sharp umbo projecting over the median hinge line. Small, subtriangular posterior auricle.

Description. – The holotype is a composite mould of the right valve of an adult specimen, with the posterior margin and the umbo slightly fractured. The specimen dimensions are given in Table 10. Moderately inflated shell with a more

Fig. 36. Rhyssomytiloides spp. from Sergipe: ×1. □ A-C. *R. mauryae* (Hessel); A. holotype (DNPM 6067): A1. lateral view, A2. anteroventral view; B. (PMU SA-173); C. (PMU SA-172); □ D-E. *R.* aff. *mauryae*; D. (PMU SA-150); E. (PMU SA-169); □ F-G. *R. bengtsoni* n. sp.; F. holotype (DNPM 6108): F1. lateral view, F2. anteroventral view; G. (PMU SA-163); □ H. *R. alatus* (Hessel), holotype (DNPM 6055): H1. dorsal view of the two valves, H2. lateral view of the right valve, H3. ventral view of the right valve; □ I. *R. beurleni* n. sp., holotype (DNPM 6106): I1. lateral view, I2. anterior view; □ J-K. *R. retirensis* n. sp.; J. holotype (DNPM 6105): J1. lateral view, J2. anteroventral view; K. (PMU SA-158): K1. dorsal view of the two valves, K2. lateral view of the left valve.

flattened posterior region. Umbo anterior, sharp, protruding, almost at the end of the hinge line. Shell ornament consists of thin, undulating, ovate, asymmetrical concentric cristae, regularly and broadly spaced. They have an average distance of 4.1 mm, up to 25 mm from the umbo (measured along the growth axis). Strong divergent ribs arranged in a more or less regular pattern (average distance of 4.8 mm), stronger in the anterior region, with broad, rounded ends. Where the ribs and cristae intersect there is a small nodule. The anterior margin is slightly curved. The ventral margin is abruptly curved in the posterior region. The posterior margin forms an obtuse angle with the hinge line. The ontogenetic changes in L/H ratio and angles are given in Fig. 35. Differentiation between juvenile and adult stages seems to be simultaneous with the appearance of the divergent ribs, at a certain ontogenetic stage. A strong umbonal fold separates the umbo from the median cardinal margin, forming a subtriangular, small posterior auricle.

Remarks on the paratypes. – The measurements of the paratypes are presented in Table 10. There are adult specimens both larger and smaller than the holotype. In PMU SA-171, a well-preserved specimen, two to three thin growth lines

Table 10. Measurements of *Rhyssomytiloides mauryae* from Retiro 26: RV – right valve; LV – left valve. Asterisk denotes inferred dimension of incomplete specimen.

Spec. No.	h	l	l/h	H	L	L/H	S	b	d
RV									
SA-149	26*	38*	1.46	34*	35*	1.03	–	09	–
SA-173	36	44*	1.22	46*	35	0.76	18	12	43°
DNPM-6109 (SA-174)	36	30	0.94	39	25	0.65	16	12	45°
LV									
SA-171	23*	27*	1.17	30*	23	0.76	–	10	–
SA-172	31*	33*	1.06	38	25*	0.65	13*	08	50°
SA-225	19	19*	1.00	23*	18*	0.78	06*	08	52°

Table 11. Measurements of *Rhyssomytiloides* aff. *mauryae* from Retiro 26 and São Roque 5: RV – right valve; LV – left valve. Asterisk denotes inferred dimension of incomplete specimen.

Spec. No.	h	l	l/h	H	L	L/H	S	b	d
RV									
SA-168	32	36*	1.12	36*	32	0.88	17?	8	56°
SA-169	34	34*	1.00	37*	27*	0.73	–	7	-
LV									
SA-150	21*	23*	1.09	25*	19	0.76	–	3	-
SA-167	32	32	1.00	39	29	0.74	18	8	48°

appear between adjacent concentric cristae. There are similar but less pronounced growth lines in specimen PMU SA-172, especially in their counterpart. The thickness of the shell is less than 0.5 mm. No internal characters are preserved. In specimen PMU SA-225 part of the articulation is visible, with four minute ligamental pits close to the umbo.

Affinities. – *Rhyssomytiloides mauryae* is different from all Turonian inoceramids described up to the present. Basse (1934, pp. 87, 93–95) mentioned from Sikily valley (southwestern Madagascar) the occurrence of *Cladoceramus* cf. *diversus*. Dr Jacques Sornay kindly provided a plaster cast of this specimen for study. The shape and ornament of the shell place it clearly in *R. mauryae*. The Madagascan specimen has both valves preserved: the right valve almost complete (only the posterior auricle is broken) and the left valve more flattened and incomplete (all margins are fractured). The right valve shows five minute ligamental pits and has the following dimensions (linear measurements in mm; * denotes inferred dimensions): h = 37; l = 49; l/h = 1.32; H = 51; L = 35; L/H = 0.68; S = 9*; b = 24; d = 53°; the left valve has the umbo poorly preserved and the following dimensions: h = 36*; l = 41*; l/h = 1.13; H = 44*; L = 34*; L/H = 0.77; S = 15*; b = 16; d = 50°.

The specimen of *Sphenoceramus* aff. *S. schmidti* figured by Klinger (1977) from an offshore drilling in South West Africa is included in *Rhyssomytiloides mauryae*, on account of its similarity in ornamentation, outline and size.

R. mauryae is different from the Cenomanian–Turonian *Inoceramus diversus* Stoliczka, 1871 (Ayyasani & Banerji 1984) in being more obliquely ovate (not subquadrangular), and in having more regular and numerous divergent ribs, a thinner shell and a posterior auricle. *I. diversus* occurs in India, where it is apparently a rare species (Stoliczka 1871).

The species which seems to be most similar to *R. mauryae* is *Inoceramus* (*Platyceramus*) *japonicus japonicus* Nagao & Matsumoto, 1940, from the Santonian of Japan and Sakhalin island, USSR (Hayami 1975). However, *R. mauryae* is smaller, more convex, obliquely elongate, with thin, well-spaced and ovate concentric cristae, whereas *I.* (*P.*) *japonicus japonicus* has a thicker shell, is less oblique, flattened with an elongate hinge line and subregular ovate concentric ornament, and has no umbonal fold between the cardinal margin and the rest of the shell (cf. Noda 1983, Pl. 41:1). *Inoceramus* (*P.*) *japonicus* is similar to *Inoceramus undulatoplicatus* Roemer, 1852, and to *Sachalinoceramus exrotivus* Glazunov, 1972, which are upper Coniacian to Santonian North Temperate species (Matsumoto 1959; Glazunov 1972). But these species are very large, slightly biconvex, less oblique and thick-shelled, resembling only slightly the Sergipe species.

Another inoceramid similar to *R. mauryae* is *Inoceramus* (*Actinoceramus*) *subsulcatiformis* Böse & Cavins, 1928, from the middle Albian of Texas. The only specimen described and illustrated (Pl. 18:1–5) has both valves, with the left valve bigger and more convex than the right valve. It is a small and inequivalve form. The concentric ornament is more flattened, the divergent ribs less numerous, the umbo much less sharp and the valves less obliquely elongate than in *R. mauryae*.

Other species of divergently ornamented inoceramids are more different from *R. mauryae*: their ornamentation is coarser and more irregular, they are generally very large and thick, they have a different umbonal region, and differ also in the presence or absence of a dorsal auricle.

R. mauryae was first described by Hessel (1986) as *Sphenoceramus mauryae*. The redescription herein is based on additional and better preserved material.

Occurrence. – Bed A at Retiro 26, Sergipe Basin; lower Turonian, in massive limestones of the Cotinguiba Formation (Laranjeiras facies), associated with *Mytiloides mytiloides*, *M. modeliaensis*, and the ammonites *Mammites* sp., *Neoptychites* sp. and *Kamerunoceras* sp. The species also occurs in Madagascar and South West Africa.

Rhyssomytiloides aff. *mauryae* (Hessel, 1986)
Fig. 36D–E

Material. – Two left and two right valves preserved as composite moulds with shell fragments. The specimen PMU SA-150 has its dorsal and anterior margins fractured. The other specimens have damaged posterior margins. PMU SA-150, 167–169.

Description. – The specimens are poorly preserved, with few morphological details preserved. The specimens are small, suboval, slightly elongate and moderately convex (Table 11). Anterior margin subrectilinear, ventral margin asymmetrically rounded, and posterior margin uniformly arcuate. Umbo prosogyrous (very close to anterior end of cardinal margin), not very sharp nor protruding. Hinge line moderately elongate, rectilinear. Posterior auricle appears very reduced. Ornamentation with asymmetrical concentric cristae, with the proximal flank steeper than the distal flank, regular in size and distance, in spite of a progressive and continuous increase of these features during ontogeny. Growth lines are not visible between adjacent cristae. There are also slightly arcuate divergent ribs with broad rounded ends. These ribs appear at a certain ontogenetic stage, between 17 and 26 mm from the umbo, measured along the growth axis. Occasionally, the divergent ribs are bifurcated. At the intersection of the ribs and cristae there is a nodule. The most significant ontogenetic modifications are the appearance of the divergent ornament and the increasing convexity. Part of the articulation is preserved in only one specimen (PMU SA-169), with six minute ligamental pits, more closely spaced close to the umbo. The internal characters are not preserved.

Remarks. – The specimens here described are different from *Rhyssomytiloides mauryae*, being more rounded (less mytilidiform), more flattened, with divergent ribs, which occasionally are bifurcated, appearing earlier in ontogeny. *R.* aff. *mauryae* also resembles the Santonian *Inoceramus (Platyceramus) japonicus japonicus* Nagao & Matsumoto, 1940, from Japan and Sakhalin (Hayami 1975, especially Fig. 8:1). However, the Sergipe specimens have a much less protruding umbo, more arcuate divergent ribs, more distant and coarser concentric cristae, and a thinner elongate shell.

Occurrence. – Bed A at Retiro 26 and at São Roque 5, Sergipe Basin; lower Turonian, in massive limestones of the Cotinguiba Formation (Laranjeiras facies), associated with *Mytiloides mytiloides*, *M. modeliaensis* and the ammonites *Mammites* sp., *Neoptychites* sp. and *Kamerunoceras* sp.

Rhyssomytiloides bengtsoni n. sp.
Figs. 36F–G, 37

Derivation of name. – In honour of the Swedish palaeontologist Peter Bengtson, who has dedicated himself to the palaeontology and stratigraphy of the Cretaceous of northeastern Brazil.

Holotype. – Specimen DNPM 6108, collected by Maria Helena Ribeiro Hessel in February 1983.

Type locality. – Retiro 26, 15 km NW of Aracaju, Sergipe, Brazil.

Type stratum. – Lower Turonian of the Cotinguiba Formation; bed A of the section here studied. The level falls within the Turonian 2 ammonite assemblage of Bengtson (1983).

Material. – Including the holotype, three left valves preserved as composite moulds, with minute shell remains near the cardinal margin or as shell fragments (PMU SA-163). The holotype is slightly damaged at the posterior margin. Specimen PMU SA-163 has the umbonal region destroyed. Specimen PMU SA-164 shows a broken ventral margin. The holotype is deposited in the Seção de Paleontologia of the Departamento Nacional da Produção Mineral (DNPM), Rio de Janeiro, under number 6108. There is a plaster cast in the Palaeontological Museum of Uppsala University, under number SA-165. The paratypes are also deposited in Uppsala, under numbers SA-163 and 227.

Table 12. Measurements of *Rhyssomytiloides bengtsoni* from Retiro 26, Retiro 17 and São Roque 5: LV – left valve. The asterisk denotes inferred dimension of incomplete specimen.

Spec. No.	h	l	l/h	H	L	L/H	S	b	d
LV									
SA-227	39*	32*	0.82	48*	28*	0.58	14*	11	38°
SA-163	38*	47	1.23	43*	40	0.93	–	10	-
DNPM-6108 (SA-165)	37	32	0.86	40*	27*	0.67	12*	11	61°

Diagnosis. – Inequilateral, ovate, thin-shelled, moderately convex, flattened in the posterior region. Poorly developed concentric cristae, almost invisible, overlapped by arcuate strong divergent ribs, subregularly spaced. Rounded and crenulated margins. Umbo prosogyrous, slightly projected over the cardinal margin, which is of medium to short length.

Description. – The holotype is a composite mould of the left valve of an adult specimen, with the posterior margin and the umbo damaged. The specimen dimensions are given in Table 12. Shell moderately convex, with the posterior region flattened and the anterior end abruptly convex. Anterior, prosogyrous umbo at the end of cardinal margin, not very sharp, nor protruding. Short and broad auricle. Anterior margin somewhat arcuate, slightly rounded up to the ventral margin. Posterior margin not preserved, except the dorsal part which forms an obtuse angle with the median hinge line. Ornament consists of flattened and regularly spaced concentric cristae (average distance of 2 mm between adjacent cristae) and of slightly arcuate, subregularly spaced divergent ribs. Ornamentation starts at 22 mm from the umbo, measured along the growth axis. Divergent ribs are less accentuated in the dorsal part. There are five ribs in the anterior region and five in the posterior. The rib ends are rounded and broad. The ontogenetic changes in angles are given in Fig. 37. The shell is thin, as shown by a fragment close to the umbo. Articulation and internal characters are not preserved.

Remarks on the paratypes. – The measurements and angles of the paratypes are summarized in Table 12. Specimen PMU SA-163 shows the thin shell preserved (thickness approxi-

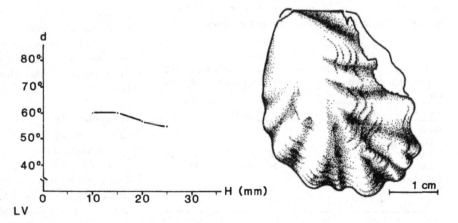

Fig. 37. Rhyssomytiloides bengtsoni n. sp. Ontogenetic variations of angle d; figure shows holotype (DNPM 6108).

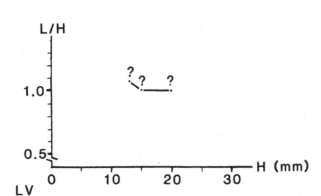
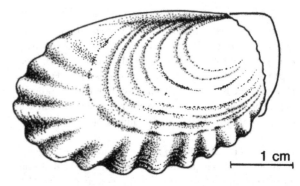

Fig. 38. Rhyssomytiloides alatus (Hessel). Ontogenetic variations of L/H ratio; figure shows holotype (DNPM 6185).

mately 0.5 mm), with its ventral and posterior margins well preserved (Fig. 36F). The general outline is more elongate than that of the holotype. Specimen PMU SA-227 is poorly preserved, except for the umbo and the articulation, and has eleven minute ligamental pits.

Affinities. – Like the other species of *Rhyssomytiloides, R. bengtsoni* shows no similarities with other previously described Turonian inoceramids, but resembles more recent ones. A species similar to *R. bengtsoni* is *Inoceramus (Platyceramus) japonicus hokkaidoensis* Noda, 1983, from the Santonian of Japan, but this subspecies of *I. (P.) japonicus* is distinctive in its thicker shell, regular concentric undulations, with growth lines and the formation of a nodule at the intersection of the divergent ribs and concentric elements. *Rhyssomytiloides bengtsoni* also resembles *I. (Cordiceramus) kanmerai* Toshimitsu, 1986, from the Santonian–Campanian of Japan (Toshimitsu 1986). However, this species is less elongate, with more delicate, divergent ornamentation, and has a more prominent, arcuate umbo than *R. bengtsoni*. From a comparison between *R. bengtsoni* and the other species of *Rhyssomytiloides* found in Sergipe, it is evident that *R. bengtsoni* has a more prominent divergent ornamentation (with arcuate and irregular development) without forming a nodule at the intersection of ribs and cristae.

Occurrence. – Bed A at Retiro 26, as well as at Retiro 17 and São Roque 5 (localities of Bengtson 1983), Sergipe Basin; lower Turonian, in massive limestones of the Cotinguiba Formation (Laranjeiras facies) associated with *Mytiloides mytiloides, M. modeliaensis, Rhyssomytiloides mauryae,* and the ammonites *Mammites* sp., *Neoptychites* sp. and *Kamerunoceras* sp.

Table 13. Measurements of *Rhyssomytiloides alatus* from Retiro 17 and São Roque 5: RV – right valve; LV – left valve. The asterisk denotes inferred dimension of incomplete specimen.

Spec. No.	h	l	l/h	H	L	L/H	S	b	d
Bivalved									
DNPM-6107 (SA-162)									
RV	26	43*	1.87	37*	26	0.63	17*	17	20°
LV	22*	39*	1.77	37*	25	0.67	14*	17*	–
Univalved									
RV									
SA-161	25*	29*	1.16	27*	25	0.92	–	12*	–

Rhyssomytiloides alatus (Hessel, 1986)

Figs. 36H, 38

Synonymy. – □ v 1986 *Sphenoceramus alatus* n. sp. – Hessel, p. 234, Pl. 1:4a–c, Textfig. 6.

Derivation of name. – From Latin *alatus*, 'wing', referring to the wing-shaped shell.

Holotype. – Specimen DNPM 6055, collected by Suzana and Peter Bengtson in 1971–1972.

Type locality. – São Roque 5, NW of Aracaju, Sergipe, Brazil.

Type stratum. – Lower Turonian of the Cotinguiba Formation, in the Turonian 2 ammonite assemblage of Bengtson (1983).

Material. – Two specimens: one with both valves (holotype) preserved as a composite mould, with the anterior portion fractured (the left valve has a more extensive fracture); the other a right valve, also as a composite mould (with shell fragments), with the anterior region not preserved. The holotype is deposited in the Seção de Paleontologia of the Departamento Nacional da Produção Mineral (DNPM), Rio de Janeiro, under number 6055. A plaster cast is kept in the Palaeontological Museum of Uppsala University, under number SA-162. The paratype is deposited in the same institution (Uppsala), under number SA-161.

Diagnosis. – Small, equivalve, inequilateral, very elongate, highly convex. Asymmetrical concentric cristae superposed by regularly disposed, divergent ribs at the shell ends, stronger in the ventral region. Median hinge line. Rounded posterior margin and subrectilinear ventral margin, both crenulate.

Description. – The holotype is a composite mould with both valves, although not well preserved: the right valve has a broken umbo, and the left valve lacks the ventral and the umbonal regions. The aperture angle is approximately 45°. The specimen measurements are given in Table 13. The shell is very inflated, with the ventral portion almost vertical and the dorsal region flattened. Posterior auricle very nar-

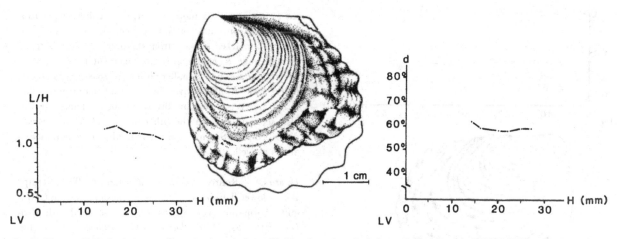

Fig. 39. Rhyssomytiloides beurleni n. sp. Ontogenetic variations of L/H ratio and angle d; figure shows holotype (DNPM 6106).

row and elongated. Ventral margin broadly rounded and strongly arcuate along the posterior margin. Ornamentation consists of regular and closely spaced, elongate-ovate, flattened asymmetric concentric cristae, superposed by divergent ribs. Average distance of the concentric cristae 2.5 mm (measured along the growth axis, between a distance of 5? to 29 mm from the umbo). The distance between the ribs varies from 3 to 5.7 mm, the ribs being very distinct at the ventral margin, less distinct at the posterior margin, and very faint at the dorsal margin. The divergent ribs start at a determined ontogenetic stage (inferred to be ca. 30 mm from the umbo, measured along the growth axis). There are eight ribs in the ventral region, four in the posterior and two in the anterior area (right valve). Rib ends are rounded. The ontogenetic changes of the L/H ratio (based on reconstructions) are given in Fig. 38. Hinge line moderately large and subrectilinear. The passage from juvenile to adult stage is not visible. Internal characters and articulation are not preserved.

Remarks on the paratypes. – There is only one paratype (PMU SA-162), which is a fragment of the posterior portion of the shell. It shows regularly spaced, asymmetrical concentric cristae bearing a small nodule at the point of intersection with the divergent ribs. The shell fragments have a thickness of less than 0.5 mm.

Affinities. – *Rhyssomytiloides alatus* differs from all known Cretaceous inoceramids, including those of Sergipe. It differs from *R. mauryae* and *R. bengtsoni* especially in its much more elongate and highly biconvex shape of the shell and its longer posterior auricle. Considering these features, as well as size, shell thickness, and details in ornamentation, it differs even more from the other divergently ornamented inoceramid known. *R. alatus* was first described by Hessel (1986) as *Sphenoceramus alatus*. The redescription herein is based on additional data.

Occurrence. – São Roque 5 and Retiro 17 (localities of Bengtson 1983) Sergipe Basin; lower Turonian, in massive limestones of the Cotinguiba Formation (Laranjeiras facies).

Rhyssomytiloides beurleni n. sp.
Figs. 36I, 39

Derivation of name. – In honour of the West German geologist and palaeontologist Karl Beurlen, who devoted the later part of his life to the geology of northeastern Brazil, producing pioneer contributions to the knowledge of macrofossils of that region.

Holotype. – Specimen DNPM 6106, collected by Peter Bengtson in January 1977.

Type locality. – Retiro 21 (Bengtson 1983), 15 km NW of Aracaju, Sergipe, Brazil.

Type stratum. – Lower Turonian of the Cotinguiba Formation, in the *Turonian 2* ammonite assemblage of Bengtson (1983).

Material. – A left valve well preserved as a composite mould, with minute shell remains. The holotype is deposited in the Seção de Paleontologia of the Departamento Nacional da Produção Mineral (DNPM), Rio de Janeiro, under number 6106. A plaster cast is kept in the Palaeontological Museum of Uppsala University, under number SA-160.

Diagnosis. – Small, thin-shelled, subequivalve, inequilateral, subrounded, moderately convex. Concentric cristae superposed by divergent rugae, with irregular orientation and distance, forming irregularly developed nodules at the shell margins. Elongate hinge line. Irregularly rounded posterior and ventral margins. Prosogyrous and slightly protruding umbo.

Table 14. Measurements of *Rhyssomytiloides beurleni* from Retiro 21: LV – left valve. Asterisk denotes inferred dimension of incomplete specimen.

Spec. No.	h	l	l/h	H	L	L/H	S	b	d
LV									
DNPM-6106 (SA-160)	35*	36	1.03	38*	37	0.97	19	14	61°

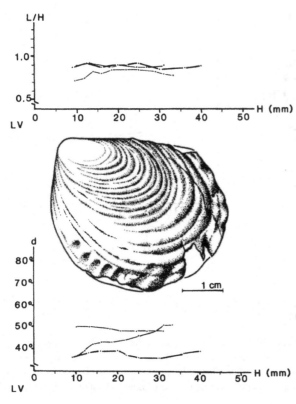

Fig. 40. Rhyssomytiloides retirensis n. sp. Ontogenetic variations of L/H ratio and angle d; figure shows holotype (DNPM 6105).

Description. – The holotype is a composite mould of the left valve of an adult specimen, with the ventral margin damaged and remains of the thin shell preserved close to the umbo. Dimensions are given in Table 14. Shell outline subcircular. Anterior margin slightly arcuate, becoming a closed curve at the ventral region. The posterior margin forms a moderate curve towards the anterior region, presenting an obtuse angle with the hinge line. The divergent pattern of ornamentation causes an irregular crenulation along the anterior, ventral and posterior margins. Umbo moderately protruding, not very sharp, prosogyrous, slightly curved, at the anterior end of the cardinal margin. The concentric ornament consists of subrounded, asymmetric cristae, initially very closely spaced and later more distant. The average distance between the cristae, measured (along the growth axis) up to 25 mm from the umbo, is 0.9. Growth lines are absent between the cristae. Divergent rugae, with very irregular arrangement, occur on the anterior, posterior and ventral margins. Irregular nodules are present at the intersection of rugae and cristae. The ontogenetic changes in L/H ratio and angles are given in Fig. 39. There is an umbonal fold near the rectilinear hinge line forming the small and elongate posterior auricle. There is no strong differentiation in the passage from the juvenile to adult stage, although the divergent rugae mark the adult stage. The juvenile shell is very compressed, and in the adult stage the shell becomes more convex. Internal characters are not preserved.

Affinities. – *Rhyssomytiloides beurleni* in its initial stage resembles in shape *Mytiloides* of the *M. hercynicus* group. However,

the former is slightly more slender, with a sharper protruding umbo, concentric ornament composed of cristae and without a marked passage from the juvenile to adult stage. In this stage *R. beurleni* presents divergent rugae, is more subcircular and has a smaller auricle. *Rhyssomytiloides beurleni* differs from *R.* aff. *mauryae*, in its broad, more protruding umbo and very irregular, divergent rugae. These features also separate it from the other species of *Rhyssomytiloides* described from Sergipe, and from all other Turonian inoceramids as well.

Occurrence. – Retiro 21 (locality of Bengtson 1983), Sergipe Basin; lower Turonian, in massive limestones of the Cotinguiba Formation (Laranjeiras facies), associated with *Mytiloides mytiloides, M. modeliaensis, Rhyssomytiloides* aff. *mauryae,* and the ammonites *Mammites* sp. and *Neoptychites* sp.

Rhyssomytiloides retirensis n. sp.
Figs. 36J–K, 40

Derivation of name. – From the farm Retiro 21, which lent its name to the locality that yielded the holotype and all the paratypes.

Holotype. – Specimen DNPM 6105, collected by Peter Bengtson in January 1977.

Type locality. – Retiro 21 (Bengtson 1983), 15 km NW of Aracaju, Sergipe, Brazil.

Type stratum. – Lower Turonian of the Cotinguiba Formation, in the *Turonian 2* ammonite assemblage of Bengtson (1983).

Material. – Including the holotype, one specimen with both valves and two left valves preserved as composite moulds, with minute shell fragments, slightly fractured in ventral or posterior margins. The holotype is deposited in the Seção de Paleontologia of the Departamento Nacional da Produção Mineral (DNPM), Rio de Janeiro, under number 6105. A plaster cast is kept in the Palaeontological Museum of Uppsala University, under number SA-159. The paratypes are in the same institution (Uppsala), under numbers SA-157 and 158.

Diagnosis. – Thin-shelled, subequivalve, inequilaterally elongate, moderately convex, with medium size. Concentric cristae irregularly distributed, superposed by irregular divergent rugae at the shell margins. Anterior, ventral and posterior margins irregularly rounded. Posterior margin forms an obtuse angle with the elongate hinge line. Umbo prosogyrous, slightly protruding, terminal.

Description. – The holotype is a composite mould of a left valve of an adult specimen with its dorsal, ventral and posterior margins slightly fractured. Dimensions are given in Table 15. The general shape of the shell is elongate-ovate, mytilidiform, with the anterior margin slightly arcuate. Ventral margin rather strongly curved at posterior region. Posterior margin moderately and uniformly rounded, forming an obtuse angle with the cardinal margin. All but the dorsal margin are slightly irregular, owing to the divergent rugae. Umbo less sharp, prosogyrous, terminal, not salient over rectilinear hinge line. Concentric ornamentation consists of ovate, irregularly developed concentric cristae, rather closely spaced on the initial shell. This ornament is completed by irregular divergent rugae, rather strong at the ventral region. Average distance of the cristae, measured up to 30 mm from the umbo (measured along the growth axis), 1.3 mm. The ontogenetic changes in L/H ratio and angles are given in Fig. 40. The passage from the juvenile to adult stage is not observable. There is a large muscle scar in the posterior region of the shell. Articulation is not preserved. Very small shell fragments close to the cardinal margin and in the posterior area suggest a thin shell.

Remarks on the paratypes. – The holotype is the most complete and well preserved specimen. Specimen PMU SA-158 presents both valves (slightly dislocated) with a thin shell fragment at the right valve. Table 15 suggests that the species is subequivalve, but it is also possible that a post-mortem flattening of one of the valves has occurred (Fig. 36). The divergent rugae occupy a larger part of the shell surface in this specimen than in the holotype. At the intersection of the concentric cristae with the divergent rugae there is a small nodule. Specimen PMU SA-157, with its entire ventral margin fractured, is more elongately ovate than the holotype. As in the holotype, a large muscle scar is visible at the posterior region of the shell.

Affinities. – *Rhyssomytiloides retirensis* differs from *R. beurleni* in being more obliquely ovate (mytilidiform), with the posterior region more flattened and a less protruding umbo. The ornamentation is rather similar to that of *R. beurleni*. *Rhyssomytiloides retirensis* is also different from the other species of *Rhyssomytiloides* from Sergipe in its irregular, divergent rugae, as well as from all other Turonian inoceramids.

Occurrence. – Retiro 21 (Bengtson 1983), Sergipe Basin; lower Turonian, in the massive limestones of the Cotinguiba Formation (Laranjeiras facies), associated with *Mytiloides mytiloides*, *M. modeliaensis*, *Rhyssomytiloides* aff. *mauryae*, *R. beurleni*, and the ammonites *Mammites* sp. and *Neoptychites* sp.

Biostratigraphy

The palaeontological succession at Retiro 26

The limestone sequence of Retiro 26 contains abundant marine macrofossils, particularly ammonites, decapod crustaceans, inoceramids and small fish scales and vertebrae. The faunal associations are listed below, bed-by-bed, employing the following scale for the relative abundance: abundant, common, relatively common, rare and absent.

In *bed A* there is an association of mytilidiform inoceramids of the *Mytiloides labiatus* group, and, in lesser numbers, divergently ornamented forms belonging in the new genus *Ryssomytiloides*. The inoceramids are in general common, mostly of medium size (3–10 cm). Abundant fish remains, decapod crustaceans and carbonized algae are associated

Table 15. Measurements of *Rhyssomytiloides retirensis* from Retiro 21: RV – right valve; LV – left valve. Asterisk denotes inferred dimension of incomplete specimen.

Spec. No.	h	l	l/h	H	L	L/H	S	b	d
Bivalved									
SA-158									
RV	40*	32*	0.80	43*	27*	0.62	–	07	-
LV	38*	36*	0.94	43*	29*	0.67	12*	09	48°
Univalved									
LV									
SA-157	26*	44*	1.04	47*	32	0.68	–	12	-
DNPM-6105	41	43	1.04	45	42	0.93	–	10	-
(SA-159)									

with this fauna; large ammonites (diameter 15–30 cm) are common (e.g., *Mammites* sp. and *Neoptychites* sp.), medium-sized and small ammonites (less than 15 cm in diameter; e.g. *Kamerunoceras* sp.) and irregular echinoids are relatively common. Macrofossils were not observed in the uppermost brecciated part of this unit.

In *bed B* inoceramids of the *M. labiatus* group are relatively common; however, specimens of *Rhyssomytiloides* are absent. As a rule, the inoceramids are medium-sized (as in the underlying bed), but some larger specimens up to 22 cm long are also found. Other organisms in this bed are: abundant fish remains, relatively common small ammonites and carbonized algae.

Bed C shows intraclastic sedimentation. Fragments of *Mytiloides* sp. are rare (in larger clasts) together with common fish scales. Some clasts are microcoquinoids or show activities of benthic organisms.

Bed D yielded no calcareous macrofossils, but fish scales and remains of crustaceans are relatively common.

Beds E to I yielded an association of subtriangular or ovate inoceramids with slightly inflated forms, belonging to the *Mytiloides hercynicus* group and *Sergipia*. In beds E and F the specimens are small (up to 3 cm) to medium sized (up to 8 cm), becoming medium to large (more than 8 cm) in the uppermost beds. *Mytiloides hercynicus* is the most abundant form. *Sergipia* increases in numbers upwards. There are abundant fish remains (teeth, scales and vertebrae), decapod crustaceans and ammonites. Of the latter, large coilopoceratids can occur in the uppermost layers. *Mammites* sp. and *Neoptychites* sp. are absent. Small, paired aptychi (6–20 mm) are relatively common and often carbonized.

Bed J yielded common *Sergipia* spp. and *Mytiloides* spp. The specimens are medium sized and associated with abundant small ammonites (*Benueites?* sp. and *Watinoceras?* sp.) and fish scales and vertebrae. There are common remains of crustaceans and relatively common minute bivalves of uncertain identity.

Bed K yielded an association quite distinct from the previous ones. Here, inoceramids, fish and crustaceans are not preserved – if they were ever present. Poorly fossilized bivalves belonging to the families Cardiidae and Poromyidae and the order Veneroida, are abundant, as are simple and branched tubular ichnofossils resulting from benthic activities; gastropods (*Pterodonta* sp.) and echinoids (*Hemiaster* cf. *jacksoni* Maury) are rare. The macrofossils are locally concentrated within the unit (Fig. 26), suggesting the existence of a bioherm-like body.

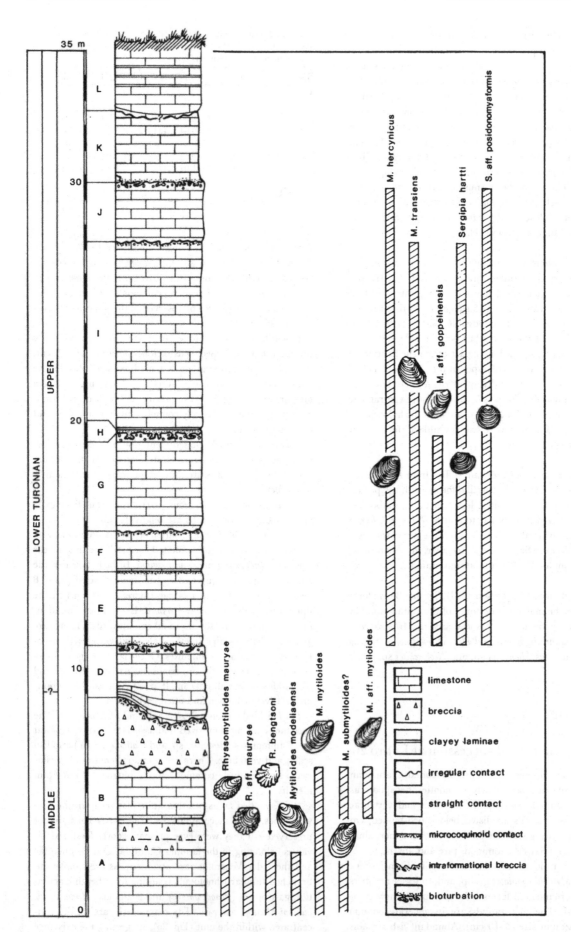

Fig. 41. Local ranges of inoceramid species in the Retiro 26 sequence.

Bed L exhibits rare fish scales, remains of crustaceans and small heteromorph ammonites in its limestone parts. Inoceramids were not found.

Figure 41 summarizes the lithostratigraphic and inoceramid data from the sequence.

Inoceramid biostratigraphy

Only 5 m of the 35 m sequence described did not yield any inoceramids (Fig. 41). These levels are those with a bioherm-like body (C and K) and the beds that overlie them (D and L). In the other units inoceramids are relatively common, although they are less common in bed E and more abundant in beds I and J. In beds B and I there is a predominance of small forms, with larger specimens present.

In *bed A, Mytiloides mytiloides* (Mantell) and *M. submytiloides?* (Seitz) occur throughout; in the lowermost part there occur *M. modeliaensis* (Sornay), *Rhyssomytiloides mauryae* (Hessel), *R. bengtsoni* and *R.* aff. *mauryae*.

Bed B yielded *M. submytiloides?, M. mytiloides* and *M.* aff. *mytiloides*, some of rather large size (approximately 20 cm).

In the brecciated *bed C* only fragments of inoceramids belonging to the *Mytiloides labiatus* group were found in some clasts.

The sequence comprising *beds E* to *I* contains a fairly uniform inoceramid succession consisting of *Mytiloides* aff. *goppelnensis* (Badillet & Sornay), *M. transiens* (Seitz), *M. hercynicus* (Petrascheck), *Sergipia* aff. *posidonomyaformis* (Maury) and *S. hartti*. In the lower part of bed G, *Sergipia* sp. was found. Bed H is very thin, and yielded only small fragments of unidentified inoceramids.

In *bed J, Sergipia* aff. *posidonomyaformis* dominates over *Mytiloides hercynicus*.

Correlations

The macrofossils of the exposures studied in the area adjacent to Retiro 26 (Fig. 6) belong chiefly in the lower faunal assemblage of the Retiro 26 sequence. The lithologies consist of hard and massive light cream limestones (Laranjeiras facies of the Sapucari Member). The localities Retiro 5, 6, 24 and 25 present clayey intercalations, Retiro 7, 23 and Ribeira 17 contain brecciated strata, and Retiro 4, 21 and Ribeira 15 are locally coquinoid. Clayey intercalations are also present in the uppermost unit of Retiro 26 but their origin here is probably the result of clay accumulation caused by weathering. The coquinoid limestone reported by Bengtson (1983) from Retiro 15 was not observed at Retiro 26. The breccia at Ribeira 17 is similar to, and probably corresponds to, that of the contact between beds I and J; this is also suggested by the altitude of the two breccia levels, when adjusted for the regional SE dip. The breccia at Retiro 23 corresponds to the uppermost part of bed A, as is strongly suggested by fragments of *M.* cf. *mytiloides* found just below the brecciated level. At Retiro 4, the breccia corresponds to either unit A or C. Inoceramids were found at 14 of the 19 localities studied: Retiro 1, 2, 3, 5, 6, 7, 8, 9, 17, 21, 22, 23, 24 and 25 (Fig. 42). The predominant forms are of the *Mytiloides labiatus* group, associated with diverse species of *Rhyssomytiloides* (from Retiro 3, 5, 6, 7, 17, 21 and 22). This association is typical of beds A and B of the Retiro 26

sequence. *M.* cf. *labiatus* were observed at two localities: Retiro 9 and 21.

Taking into account the other localities described by Bengtson (1983), and also unpublished biostratigraphical data provided by him, the lowermost beds at Retiro 26 can be correlated with the locality São Roque 5. This is indicated by *Rhyssomytiloides* spp. and ammonites of the genera *Neoptychites, Mammites* and *Kamerunoceras*, and of the family Coilopoceratidae. Retiro 26 can also be correlated with Recreio 1, São Roque 4, Bumburum 4 and São Pedro 1 (Bengtson, 1983) through the occurrence of the same ammonite association. The localities Ribeira 7 and 12 can be correlated with units E to J at Retiro 26, as indicated by *Sergipia* spp., *Neoptychites* sp., *Kamerunoceras* sp. and coilopoceratids.

Correlation of the inoceramid succession of Retiro 26 with other Brazilian occurrences, outside Sergipe, is severely limited by the lack of detailed palaeontological and stratigraphical data from the latter areas. *Mytiloides submytiloides* occurs in the Jandaíra Formation (Buraco d'Agua, Açu River valley, Rio Grande do Norte: K. Beurlen 1961), originally reported as *Inoceramus labiatus*. My examination of the specimen (No. 717), kept at the Departamento de Geologia of the Universidade Federal de Pernambuco has shown it to be a *M. submytiloides*, characterized by an obliquely elongate shell, moderately inflated, with a slightly protruding umbo, subregular cristae without growth lines in between and a subtriangular auricle. It is therefore possible to establish a biostratigraphical correlation between the lowermost beds (A and B) of the Cotinguiba Formation exposed at Retiro 26 and the Jandaíra limestones at Buraco d'Agua. Additional data are, however, necessary to reinforce this correlation.

Mytiloides modeliaensis and *Sergipia* sp. are known from other localities in South America. They occur in northern Colombia (Heinz 1928a; Sornay 1981), associated with *M. labiatus* in the lower Turonian of the La Frontera Formation (Olsson 1956; Etayo-Serna 1964). Sornay (1981), using data by Reeside (1928), recognized *M. modeliaensis* in Ecuador. In Peru (Willard 1966) and Venezuela (La Luna Formation, according to Rutsch & Salvador 1954) there are some occurrences of *Sergipia*. Heinz (1928a) described *Inoceramus plicatus* var. *hercynica* from Venezuela. Correlation with other parts of South America is hampered by the same lack of information as for the Brazilian strata. However, possible correlation between the Cotinguiba Formation and La Frontera and La Luna formations can be inferred from available data.

As regards Africa, the greatest similarity is found with the Ezeaku Formation in the middle Benue valley of Nigeria. The occurrence of *Mytiloides hercynicus, M. transiens?* and *Sergipia hartti* [=*Sergipia* aff. *posidonomyaformis*] (Offodile 1976; Offodile & Reyment 1977), in combination with the presence of similar ammonites on both sides of the South Atlantic, confirms the previously known possibility of a correlation between the Cotinguiba and Ezeaku formations (Offodile 1976). In the Turonian of Nigeria, at Aka-Eze, Reyment (1955) also described ten specimens as *I.* aff. *labiatus* Schlotheim and illustrated one (Pl. 3:7). This specimen can be referred to *M. submytiloides* on account of its general shape and ornamentation. Sornay (1981) mentioned the affinity between *M. modeliaensis* from Colombia and a form found in the Tarfaya Basin, Morocco. The occurrence of *Rhyssomytiloides mauryae* in Madagascar and South West Africa (see p.

LOCALITIES	INOCERAMIDS	AMMONITES	OTHER MACROFOSSILS
Retiro 1	M. cf. modeliaensis	---	---
Retiro 2	M. cf. submytiloides	---	crustaceans echinoids fish scales
Retiro 3	M. cf. modeliaensis M. cf. mytiloides M. cf. submytiloides Rhyssomytiloides sp.	Neoptychites sp. Watinoceras sp.	echinoids fish scales gastropods
Retiro 4	---	coilopoceratids	---
Retiro 5	M. cf. mytiloides Rhyssomytiloides sp.	---	fish
Retiro 6	M. cf. mytiloides M. cf. submytiloides Rhyssomytiloides sp.	coilopoceratids Mammites sp.	echinoids
Retiro 7	M. cf. modeliaensis M. cf. mytiloides Rhyssomytilodes sp.	---	echinoids
Retiro 8	M. cf. modeliaensis M. cf. mytiloides M. cf. submytiloides Mytiloides sp.	Kamerunoceras? sp. Mammites sp. Neoptychites? sp.	fish gastropods
Retiro 9	M. cf. labiatus M. cf. mytiloides M. cf. modeliaensis	---	---
Retiro 17	M. cf. mytiloides R. alatus R. bengtsoni	coilopoceratids Watinoceras? sp	echinoids
Retiro 18	---	indet.	fish scales
Retiro 21	M. cf. labiatus M. cf. modeliaensis M. cf. mytiloides M. cf. submytiloides R. beurleni R. aff. mauryae R. retirensis	Mammites sp. Neoptychites sp. Watinoceras sp.	echinoids
Retiro 22	M. cf. modeliaensis M. cf. mytiloides Rhyssomytiloides sp.	indet.	echinoids fish scales
Retiro 23	M. cf. mytiloides	---	crustaceans fish scales non-inoceramid bivalves
Retiro 24	M. cf. mytiloides	---	echinoids
Retiro 25	indet.	indet.	---
Ribeira 15	---	indet.	non-inoceramid bivalves
Ribeira 16	---	coilopoceratids	---
Ribeira 17	---	---	---

Fig. 42. Macrofossils of the localities adjacent to Retiro 26 (ammonite data from P. Bengtson, personal communication 1986).

		ASSOCIATIONS	INOCERAMIDS	AMMONITES
LOWER TURONIAN	UPPER	Mytiloides hercynicus	M. aff. goppelnensis M. hercynicus M. transiens Sergipia hartti S. aff. posidonomyaformis	Benueites sp. coilopoceratids
		?		
		?		
	MIDDLE	Mytiloides mytiloides	M. modeliaensis M. mytiloides M. submytiloides Rhyssomytiloides mauryae R. aff. mauryae R. bengtsoni R. beurleni R. retirensis	Kamerunoceras sp. Mammites sp. Neoptychites sp. Watinoceras sp.

Fig. 43. Subdivision of the Retiro 26 sequence based on inoceramids, with the corresponding ammonite and inoceramid faunas. Boundary between middle and upper lower Turonian inferred from Tröger (1981) and Sornay (1986).

28) is also very interesting for correlation studies. However, more data on the inoceramids from the Southern Hemisphere are necessary to allow for more accurate correlations; this applies also to the occurrences of Turonian inoceramids in Gabon and Cameroun (Riedel 1932; Dartevelle & Freneix 1957) and Angola (Cooper 1978).

In the Turonian of the Antarctic the forms so far recognized belong to species other than those present at Retiro 26 (Crame 1981, 1982). All the Antarctic forms lack divergent ornamentation. The Turonian of New Zealand yields *I. mytiloides* (Wellman 1959); and inoceramids of the *I. labiatus* group are known from South India, together with *Inoceramus diversus*, a divergently ornamented species (Stoliczka 1871, Ayyasami & Banerji 1984).

Similar assemblages to those at Retiro 26 are also found in the Turonian to lower Coniacian of the Northern Hemisphere, particularly in Central America and the Caribbean. In the Dutch West Indies, Kauffman & Sohl (1978) reported the presence of *M. transiens* and *M. hercynicus*. The latter species is also known from Trinidad (Kauffman 1978a) and Mexico (Böse & Cavins 1928) associated with *Sergipia* sp. (Kauffman 1977b). Besides the three forms already mentioned, *M. modeliaensis* is also reported by Sornay (1981) from Mexico. *Sergipia* sp. is also reported by Kauffman (1977b) from California and Japan. Kauffman's current review of this genus should contribute to the clarification of possible correlations. However, it is already clear that there are similarities between the Turonian inoceramid faunas of Sergipe and Mexico, with the similar occurrence of four non-endemic species.

Rhyssomytiloides is described here for the first time from Sergipe. Divergently ornamented inoceramids of different

ages have previously been described from other areas, e.g., Ecuador (Marks 1956), Colombia, Peru and Venezuela (Heinz 1928a), Mexico (Böse & Cavins 1928), northern and western Europe (Kauffman 1979), Gabon and Madagascar (Heinz 1932) and South Africa (Matsumoto 1959; Kennedy & Klinger 1975). However, these species differ considerably from those of Sergipe in their shape, thickness, ornamentation and size of the valves. They resemble more closely *Inoceramus (Platyceramus)* from the Santonian to Campanian of Japan and the northern Pacific coast (cf. Pergament & Tröger 1979). Nevertheless, in south-western Madagascar, *Rhyssomytiloides mauryae* is found in the lower Turonian with *M. labiatus*. *R. mauryae* occurs also in South West Africa. It is therefore possible that *Rhyssomytiloides* also existed in other parts of the Southern Hemisphere during the mid-Cretaceous.

In summary, there are broad similarities between the inoceramid fauna of Retiro 26 and those of Colombia, Mexico and Nigeria, but further studies are necessary for more detailed correlations.

Inoceramid subdivision of Retiro 26

Two distinct associations of inoceramids can be recognized in the Retiro 26 area. The subdivision based on these associations is here put forward as a working hypothesis for a future biozonation, which must by necessity be based on studies of a larger geographical area (Fig. 43).

There seems to be a consensus in the current literature that *Mytiloides mytiloides* occurs in the lower Turonian all over the world. Among others, Tröger (1981) and Kauffman et al. (1978) mention *M. submytiloides* at the base of the lower

Turonian. However, the occurrence of *M. hercynicus* is a little controversial. In Portugal (Berthou & Lauverjat 1978), Romania (Lupu 1978), Venezuela, Peru, northern Poland, the Russian platform, the Caucasus (Seibertz 1979) and Germany (Tröger 1981) this species is said to occur in the uppermost lower Turonian. In southern England and France (Seibertz 1979), in Germany (Keller 1982), in northern Spain (Wiedmann & Kauffman 1978), Czechoslovakia (Kauffman 1978c), Caribbean (Kauffman 1978a) and in the Western Interior of North America (Kauffman 1977b; Kauffman *et al.* 1978), *M. hercynicus* is referred to the base of the middle Turonian. According to Tröger (1981), the boundary between the lower and the middle Turonian should be placed at the appearance of *Inoceramus apicalis* Woods, 1912. Inoceramids of the *I. lamarcki* group, to which *I. apicalis* belongs, are considered to be typical of the middle Turonian by most authors (e.g., Seitz 1959; Wiedmann & Kauffman 1978; Seibertz 1979; Tröger 1981; Keller 1982; Robaszynski 1982). There is some doubt also about the first appearance of *M. transiens*: some authors place it in the upper part of the lower Turonian (e.g., in Japan, cf. Seibertz 1979; in the Western Interior of North America, cf. Kauffman 1977b; and Bonaire, cf. Kauffman & Sohl 1978); others have suggested that it appears at the base of the middle Turonian: Tarfaya (Seibertz 1979), the Caribbean (Kauffman 1978a) and northern Spain (Wiedmann & Kauffman 1978).

In the present study the inoceramid and ammonite subdivision of Tröger (1981) and Sornay (1986) is adopted. Following this, the Retiro 26 fauna is assigned to the lower Turonian, indicated by the inoceramids *Mytiloides mytiloides*, *M. submytiloides?*, *M. hercynicus* and *M. transiens*, by the co-occurring ammonite genera *Mammites*, *Neoptychites*, *Benueites?*, and by the absence of forms of the *Inoceramus lamarcki* group and of typically middle Turonian ammonites (Bengtson, personal communication, 1986). The entire sequence at Retiro 26 can be assigned to the ammonite assemblage Turonian 2 of Bengtson (1983), which is dominated by *Neoptychites*, *Mammites*, *Kamerunoceras*, *Watinoceras* and coilopoceratids.

The recent proposal by Asmus & Campos (1983) that the lower Turonian in Brazil should be recognized by the *Pseudaspidoceras–Vascoceras–Inoceramus labiatus* biozone, does not seem to be viable, as this species (named *Mytiloides labiatus* by Kauffman & Powell 1977) is not nearly as common in Brazil as *Mytiloides mytiloides*. In the sequence at Retiro 26 *M. labiatus* was not found, and its presence could not be confirmed in the adjacent localities either.

At Retiro 26, *Mytiloides mytiloides*, *M. submytiloides?* and *Rhyssomytiloides* spp. are older than *M. hercynicus*, *M. transiens*, *M.* aff. *goppelnensis* and *Sergipia* spp. It seems quite evident that the latter species occupied the ecological niche left available by the former.

The time span of the sequence exposed at Retiro 26 is somewhat difficult to establish. Available data about the duration of inoceramid species are still insufficient and/or controversial. According to Kauffman (1979), based on radiometric dating of bentonite beds in the Western Interior of North America, *Mytiloides mytiloides* had an approximate duration of 0.4 Ma and *M. hercynicus* of 0.1 Ma, with an interval of 0.2 Ma between the disappearance of the former and the appearance of the latter. The lowest level of *M. mytiloides* at

Retiro 26 probably occurs below the sequence exposed. Also, there are nearly 5 m of limestones deposited after the disappearance of *M. hercynicus*. The duration of the Turonian Age has been calculated between 1.5 and 3 Ma (e.g. Harland *et al.* 1982; Hallam *et al.* 1985; Odin 1985; and Haq *et al.* 1987). Assuming a nearly instantaneous dispersal of the inoceramids, Kauffman's (1979) values should be applicable to the Sergipe inoceramids as well. This gives an estimated time span of 0.2–0.4 Ma for the Retiro 26 sequence. This estimate is based on the fact that beds with *M. hercynicus* occupy approximately 18 m in the sequence, and those containing *M. mytiloides* comprise 6 m. The two occurrences are separated by approximately 5 m containing a breccia bed. The estimated time span would imply an *average* sedimentation rate of 15–20 cm per thousand years. The numerous breaks in sedimentation suggest a higher actual sedimentation rate (perhaps in the order of up to 30–40 cm/1000 years), as part of the material was eroded away.

Palaeoecology and geological history

The lithologies and fossil associations of the Retiro 26 sequence indicate an outer carbonate platform environment. Ten stratigraphical breaks can be identified and suggest submarine erosion by currents. An overall picture of the geological history of the site is given in Fig. 44.

Based on the detailed analysis of the fossil associations and the lithological variations described below it is possible to reconstruct the palaeoenvironmental history of Retiro 26. In summary, the *first phase* (beds A and B) initially involved quiet sedimentation on a soft substratum in a shallow-water marine environment, as is suggested by the abundant fauna of fish, crustaceans, echinoids, ammonites, algae and both semi-infaunal and epifaunal inoceramids. During the *second phase* (bed C), when the calcareous breccia was formed, the waters must have been even shallower, agitated and well oxygenated allowing for the presence of tubular ichnofossils in bioherm-like bodies. The *third phase* (beds D to J), corresponds to the major part of the sequence. It seems to have started with ecological restrictions, followed by erosion and the formation of omission surfaces. The fauna of ammonites, fish, crustaceans and semi-infaunal inoceramids indicate an environment with a soft bottom under euhaline shallow waters, intermittently stirred and with oxygen levels slightly below normal. In the *fourth phase* (bed K) the fauna preserved is restricted to echinoids, small bivalves, gastropods and tubular ichnofossils indicating well oxygenated, shallow waters. Despite deep weathering of the upper part of the

---

Fig. 44. Fossil assemblages, lithostratigraphy and palaeoenvironmental interpretation of the sequence at Retiro 26. The symbols denote:

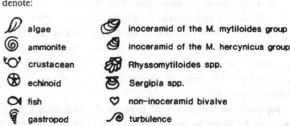

algae
ammonite
crustacean
echinoid
fish
gastropod

inoceramid of the M. mytiloides group
inoceramid of the M. hercynicus group
Rhyssomytiloides spp.
Sergipia spp.
non-inoceramid bivalve
turbulence

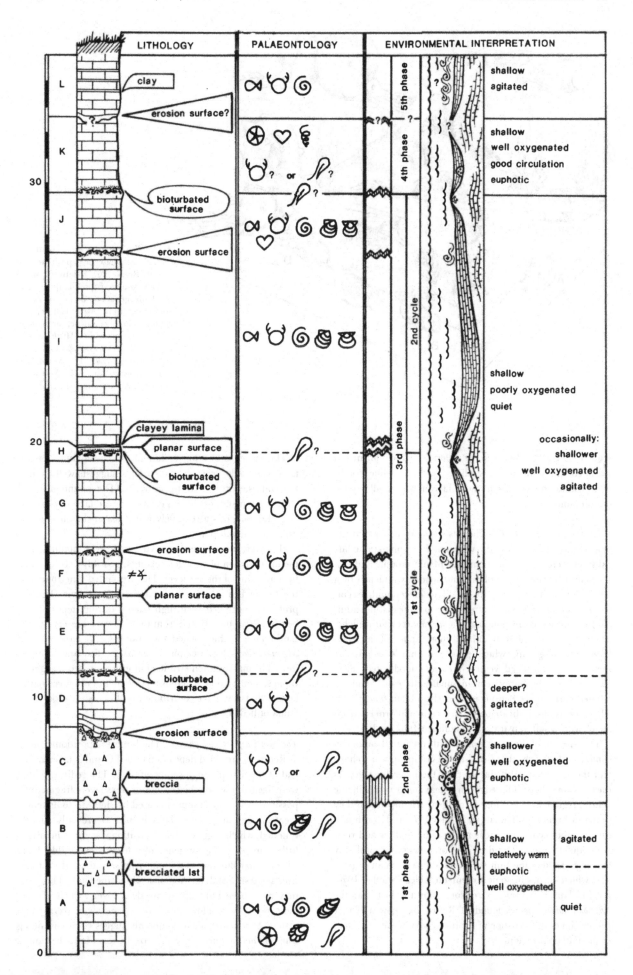

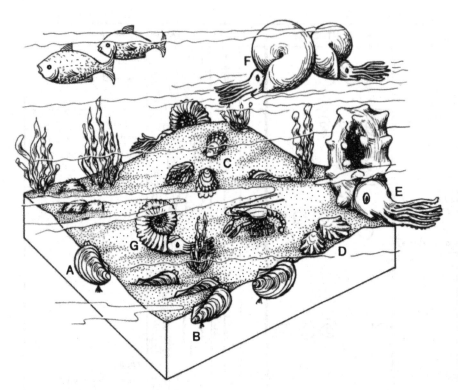

Fig. 45. Environmental reconstruction of the first phase of deposition (bed A) at Retiro 26. Relatively warm shallow waters, euphotic, quiet, with good oxygenation and circulation; soft substratum, with fish, crustaceans, echinoids, algae, ammonites and inoceramids. □ A. *Mytiloides mytiloides*. □ B. *M. submytiloides*. □ C. *Rhyssomytiloides mauryae*. □ D. *R. bengtsoni*. □ E. *Mammites* sp.. □ F. *Neoptychites* sp.. □ G. *Kamerunoceras* sp.

sequence, a *fifth phase* can be inferred, which formed a stratified limestone under shallow water, euhaline conditions, containing fish remains, crustaceans and small heteromorph ammonites.

The first depositional phase. – There is strong evidence that the inoceramids that possessed a very long-lived planktonic larval stage were eurytopic organisms, planktotrophic and sessile benthic in their adult phase, and generally attached by a byssus (Kauffman 1975, 1977a). The inoceramid assemblage found in beds A and B records two quite distinct adult life habits, each corresponding to a specific ecological niche. The elongate and slightly inflated inoceramids of the *Mytiloides labiatus* group, which possessed thin subequivalve shells with a flattened ventral area and moderately sized auricles, were probably semi-infaunal and sessile, and inhabited current-free benthic regions (Kauffman *et al.* 1977). These forms also indicate outer carbonate platform environment, with a soft substratum (Kauffman 1967, p. 104).

The species of *Rhyssomytiloides*, with an initial ontogeny similar to that of the *Mytiloides labiatus* group, appear to have had the same life habit during their the juvenile stage. As they became more inflated, with a maximum width in the ventral region, and developed divergent ornamentation they probably became epifaunal (cf. Stanley 1970, p. 173), inhabiting another ecological niche in the same shallow and relatively quiet waters. The divergent ornamentation probably helped to protect the thin shell from damage, in analogy to that which occurs in some Recent Pteroidea (Stanley 1970, p. 29). This hypothesis is reinforced by the fact that the *Rhyssomytiloides* species found at Retiro 26, show a higher proportion of preservation of the distal part of the shell than of the umbonal region.

The presence of other epifaunal or semi-infaunal organisms (echinoids, crustaceans) strongly suggests well-circulated waters, without strong currents. The abundance and variety of macrofossils, especially the epifaunal elements (Fig. 45), suggests a relatively warm sea at the onset of the first depositional phase.

However, this favourable environment was modified with the introduction of stronger currents, possibly accompanied by a lowering of the sea level. The brecciated limestone near the top of bed A reflects this new environment. The sea probably continued with high energy during deposition of bed B, as can be deduced from the exclusive preservation of semi-infaunal, thin-shelled inoceramids (*M. mytiloides* and *M. submytiloides*) or nektonic organisms (fish and ammonites). The massive, non-stratified limestone of bed B suggests rapid sedimentation. There is no evidence of other ecological factors that may have contributed to the modification of the fauna of bed A.

The second depositional phase. – The largest depositional break at Retiro 26 preceded deposition of bed C and is taken as the end of the first phase of sedimentation. The sediments that were formed towards the end of this phase were fragmented, possibly by a local slumping caused by slight tectonic movements in the area. The clasts in bed C are only slightly rounded, which suggests that deposition was rapid involving little transport. The presence of a local bioherm-like body with tubular burrows of no longer preserved benthic organisms suggests a stabilization of a submarine bank. This bank may have been built up by an algal cover, which in itself indicates very shallow (50–100 m) and well oxygenated waters. *Stromatactis*, algae commonly suggested as sustaining some bioherms during the Palaeozoic, form branched tubes

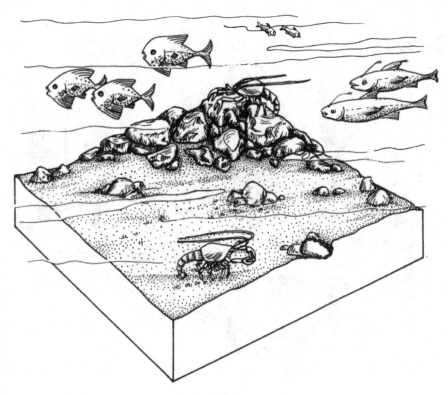

Fig. 46. Environmental reconstruction of the beginning of the third depositional phase (bed D) at Retiro 26. There were probably ecological restrictions (higher energy? or greater depth?) to benthic life, because the sea seems to have been inhabited only by crustaceans and fish.

similar to the galleries of burrowing animals (Juignet & Kennedy 1974). Algal rather than invertebrate activity may have been responsible for the tubular structures similar to *Thalassinoides* seen in Figs. 12 and 13. Juignet & Kennedy (1974) suggest that chert nodules (Fig. 15) probably form by replacement mechanisms, and that the process is initiated in heterogeneous areas of the sediment, which are generally related to bioturbated horizons. Berthou & Bengtson (1988) mentioned as a conspicuous feature in the Cotinguiba Formation the abundance of silicified fossil fragments. They suggest that geochemical conditions related to the proximity of the expanding Atlantic Ocean floor and/or great amounts of radiolarians and siliceous sponge spicules may have been responsible for this phenomenon.

The third depositional phase. – The inoceramids found in beds E through J of the third phase of deposition belong to the *Mytiloides hercynicus* group and to *Sergipia*. They show similar life habits; the specimens of *Mytiloides* with subtriangular-ovate, subequilateral, and moderately inflated thin shells, corresponding to the adult semi-infaunal forms, are typical of shallow waters and a soft substrate of variable grain size (Kauffman 1969). The *Sergipia* specimens, which also possessed thin shells and subequilateral valves, were, on the other hand, less inflated, subcircular and may have had an anterior auricle. These morphological characters point to a semi-infaunal, weakly byssate mode of life. These bivalves probably inhabited shallow waters of significant but intermittent current activity. Such an environment is also suggested by the numerous erosion and bioturbed surfaces that occur in the beds containing *Mytiloides* of the *M. hercynicus* group and *Sergipia* spp. (Fig. 44). According to Kennedy & Garrison (1975), such erosion surfaces indicate an outer

neritic environment influenced by wave activity where water depth varies from 50 to 100 m. The presence of crustacean remains throughout the sequence at Retiro 26 may suggest that these animals were responsible for a large part of the bioturbation on the erosion surfaces (Figs. 18 and 21). The occurrence of microcoquinoid and carbonized material, particularly in the lowermost horizons of each bed, immediately overlying the erosion surfaces (Figs. 18 and 20), suggests a shallowing of the sea and/or submarine erosion.

At the onset of the third phase of sedimentation (bed D), there appear to have been ecological constraints on benthic life. Only crustacean and fish remains are preserved (Fig. 46). The lack of inoceramids may have been due to the waters being very agitated or very deep, as small variations in temperature or oxygen content do not usually affect the more tolerant inoceramid faunas (Kauffman 1975). The influx of terrigenous material was low or non-existent. Possible salinity, turbidity and bottom changes are difficult to assess. An alternative to an ecological explanation for the absence of inoceramids is that shells were never preserved or were destroyed during diagenesis.

The formation of an erosion surface showing bioturbation immediately above in the sequence points to favourable oxygen conditions and the development of a bottom life. From this point, the sedimentation cycles described are limited by erosion surfaces with bioturbation (Fig. 44).

The first cycle of the third phase corresponds to beds E, F and G and the second cycle to beds H, I and J (Fig. 44). Between the biogenic erosion surfaces that limit these cycles (between D and E, G and H, and J and K), there are planar omission surfaces (between E and F, and H and I), followed by an irregular erosion surface, with an intraformational breccia (between F and G, and I and J). However, through-

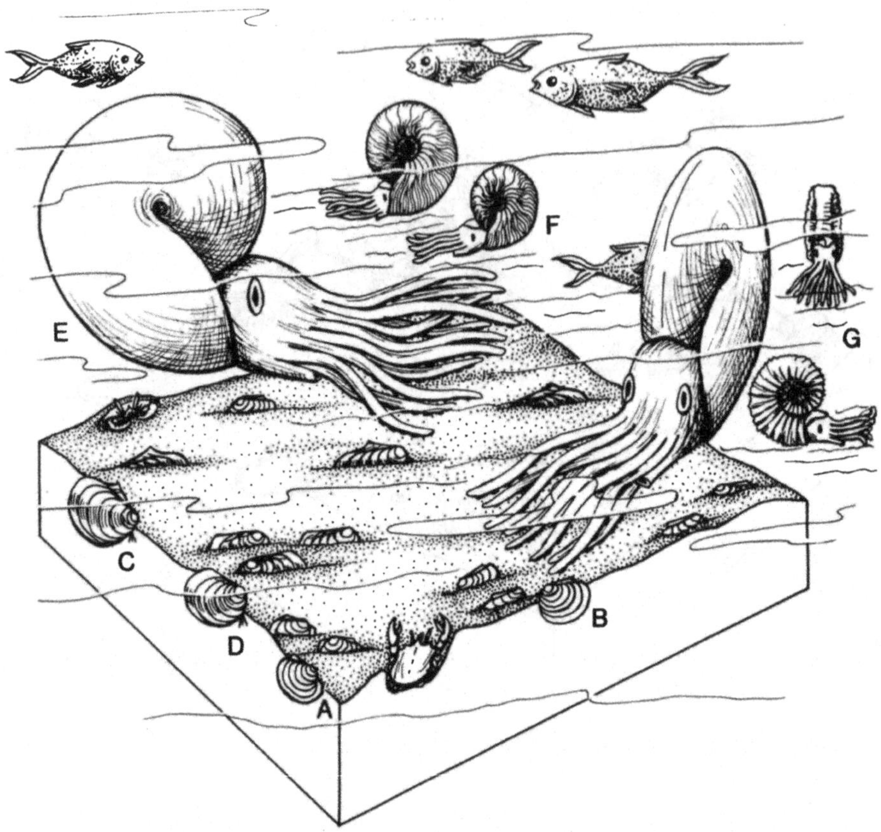

Fig. 47. Environmental reconstruction of the 2nd cycle of the third depositional phase (bed I) at Retiro 26. Shallow waters, occasionally poorly oxygenated. Soft bottom with crustaceans, fish, inoceramids and ammonites. □A. *Sergipia* aff. *posidonomyaformis.* □B. *S. hartti.* □C. *Mytiloides hercynicus.* □D. *M. transiens.* □E. coilopoceratid. □F. *Benueites?* sp.. □G. *Watinoceras* sp.

out the sequence the fauna is very uniform, indicating that the environmental variations were minor and predominantly affected by bathymetric oscillations and improved water circulation.

Once a surface with perforations of the *Thalassinoides* type had been formed in well oxygenated and very shallow waters, the sedimentary cycle started with erosion of this surface, through intense current activity. With the subsequent rise of sea level, a favourable, quiet environment was developed where ammonites, fish, crustaceans and inoceramids dominated the fauna. As a result of shallower waters and/or strong currents a second erosional event occured, followed by formation of microcoquinas. When the water level rose, the previous environmental conditions returned. With the subsequent shallowing and/or the introduction of strong currents, intraformational breccias were formed, containing clasts with bleached halos (Fig. 21). By the end of this cycle, previous conditions were reestablished, until a new erosion surface with bioturbation was formed. This was the beginning of the second cycle of the third phase, during which the processes described were repeated.

In the second cycle a clayey lamina occurs above the planar omission surface between beds H and I (Fig. 44). This intercalation indicates a sudden and short-lived local(?) increase of terrigenous matter in relation to the carbonatic production.

Throughout the third phase the oxygen level was lower than normal near the sea bottom. The absence of echinoids, gastropods, other bivalves and many other benthic organisms suggests that the bottom waters were not well oxygenated (Fig. 47), a fact that did not affect the tolerant inoceramids (according to Kauffman 1975). However, the biotur-

bated surfaces formed throughout the depositional phase could indicate tendencies to normalization of the oxygen level.

The fourth depositional phase. – In the fourth phase of deposition at Retiro 26, inoceramid, ammonite, fish or crustacean remains are absent. The presence of irregular echinoids indicates euhaline waters. Biogenic perforations of a possible bioherm-like body suggest shallow and well-oxygenated waters, and small epifaunal or semi-infaunal bivalves and gastropods, suggest a mixed soft and solid substratum (Fig. 48).

A possible fifth depositional phase. – It is probable that a lowering of the sea level followed the fourth depositional phase, but any evidence has been obscured by weathering. The overlying bed L suggests an environment perhaps similar to the third phase, with fish, crustaceans and ammonites. The absence of inoceramids may be explained as a result of high water energy, as in bed D (see p. 41).

Summary. – It is possible that the shallow, relatively warm and quiet sea, which prevailed during the initial stages of deposition, became more agitated, and possibly subject to bathymetric variations. Shortly thereafter, high-energy currents appeared, followed by local slumping. The water depth and circulation oscillated; during shallower episodes the sea was agitated, well-oxygenated, and during deeper episodes it was quiet with possible oxygen restrictions to benthic life. The environment was outer carbonate platform, with open sea circulation. Finally, the sea was shallow with normal oxygenation.

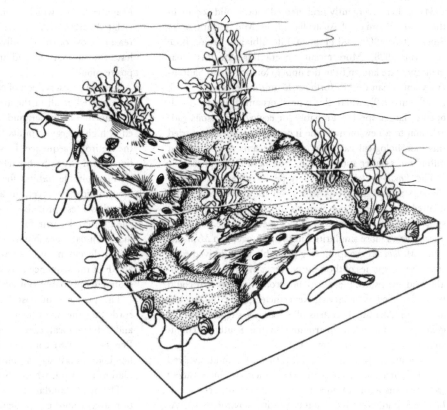

Fig. 48. Environmental reconstruction of the fourth depositional phase at Retiro 26. Well-oxygenated and shallow marine waters on/in a mixed substratum, with bivalves, gastropods, crustaceans(?), echinoids and algae(?).

Discussion

It is generally assumed that the climate during the Cretaceous was significantly warmer and more uniform than the present climate, and that during the Cretaceous there were less pronounced seasons. There was also less temperature variation between different latitudes and water depths. According to Hallam (1981), the tropical and subtropical climatic belt of the mid-Cretaceous extended to 45° latitude N and S. The inoceramid fauna of Retiro 26 does not contradict these climatic assumptions. It is possible that the Sergipe region tended to a subtropical rather than tropical climate during the early Turonian, considering the modest dimensions of the inoceramids and their thin shells. Warm water is considered the most important single factor that causes large dimensions in Recent bivalves (Nicol 1964). The absence of corals and rudists, which are abundant in the Tethyan regions, confirms this hypothesis. It can therefore be assumed that the South Temperate Subprovince (Kauffman 1973) included both subtropical and temperate climatic zones during the early Turonian.

The epicontinental sea of the initial deposition at Retiro 26 was quiet, containing some endemic species. This suggests the presence of a dispersal barrier. Reyment *et al.* (1976) put forward the hypothesis of a structural high (horst?) during the Turonian–Coniacian east of Aracaju, to explain the absence of fossils of this time interval on the continental shelf. Such a high, if it existed, would also have blocked off oceanic circulation in the middle part of the early Turonian. However, Berthou & Bengtson (1988) favour another model for these anomalies. The problem needs further study.

During the time interval corresponding to the larger part of the sequence at Retiro 26 (late early Turonian), the sea became increasingly more open, allowing good superficial circulation. The inoceramid fauna adds further support to this interpretation, by showing a decreasing percentage of endemic species in the upper part of the sequence. It would be interesting to study in detail the other co-occurring organisms (particularly the ammonites and microfossils) to either confirm or discard this hypothesis.

The presence of several unconformities in the sequence suggests that during the early Turonian the water depth oscillated significantly. Thus, there were short-lived regressive episodes throughout the deposition of the Cotinguiba Formation. These regional (or local?) episodes could also have produced a shallow-water environment on the outer continental shelf (cf. Asmus & Guazelli 1981; Berthou & Bengtson 1988) This agrees with Ojeda & Fugita (1976, Fig. 18), who placed the region of Retiro 26 between the 50 to 200 m isobath during this period. Bengtson (1983) mentioned the occurrence in the Cotinguiba Formation of several 'omission surfaces, nodular limestones, intraformational conglomerates and incipient hardgrounds', while emphasizing the discernible stratigraphical discontinuities in this formation. Fully developed hardgrounds do not occur at Retiro 26. At the time of Bengtson's visit to the quarry Retiro 26 (then Retiro 15), only 20 m of the sequence were exposed and only five of the ten discontinuity surfaces described herein were observed.

The *Rhyssomytiloides* spp. constitute the first record of inoceramids with regular, divergent ornamentation from the lower Turonian of South America. Divergently ornamented inoceramids are reported from the Cenomanian–Turonian of New Zealand (Raine *et al.* 1981), India (Stoliczka 1871), South West Africa (Klinger 1977) and Madagascar (Basse

1934). Other divergently ornamented inoceramids occur in the Lower Permian of Australia, in the Middle Jurassic of Siberia (Cox 1969) and in the middle Albian of Texas (Böse & Cavins 1928). More recent inoceramids with this ornamentation are known from the upper Coniacian and Santonian–Campanian of the Northern Hemisphere (Tröger 1976; Kauffman 1979), where this type of ornament is well developed in larger species. Probably for ecological reasons (adaptation to a new niche?), the inoceramids repeatedly tried this morphological variation, which became more common initially in smaller species in the Southern Hemisphere.

The *Rhyssomytiloides* from Sergipe is rather different in shape, size and shell thickness from the divergently ornamented Cretaceous inoceramids of the Northern Hemisphere and belong to another phylogenetic lineage. *Rhyssomytiloides* shows more similarities in their ontogeny to *Mytiloides* of the *M. labiatus* group, having probably been derived from them. Some species of the Sergipe *Rhyssomytiloides* were probably endemic, since there were no strong currents that could have dispersed their larvae. Nevertheless, the Madagascan and South African specimens of *R. mauryae* suggest that the genus is more widely distributed in the Southern Hemisphere than previously thought.

According to Kauffman (1977b), the inoceramids evolved rapidly, chiefly as a consequence of their trophic strategy (selective suspension-feeders), life habit (usually sessile epibenthos) and preferred habitat (neritic environment). At Retiro 26, *M. mytiloides* and *M. submytiloides*? occur lower in the sequence than *M.* aff. *goppelnensis* and *M. hercynicus*. This stratigraphic behaviour was reported also by Kauffman (1977b, 1978a), Seibertz (1979) and Tröger (1981). However, Seitz (1935), Sornay (1982) and Keller (1982) reported the coexistence in the Turonian type-area of *M. mytiloides*, *M. transiens*, *M. goppelnensis* and *M. hercynicus*. The true ranges of these species in various parts of the world are yet to be investigated in detail.

The co-occurrence of *Sergipia* aff. *posidonomyaformis* and *S. hartti* does not allow the identification of their phylogenetic relationship, because of the stratigraphic break that precedes their occurrences at Retiro 26. Neither is it possible to draw any further conclusions on the probable phylogenetic line from *Sergipia* to *Didymotis*, suggested by their shapes, shell thickness, ornamentation (especially considering the Nigerian specimen of *S. hartti*) and stratigraphical occurrences. *Didymotis* is common in Coniacian strata of the Cotinguiba Formation (Kauffman & Bengtson 1985).

According to Kauffman (1979), the principal environmental factor which affected the benthic invertebrate fauna during the Cretaceous was the dissolved oxygen level, which was lower in the Cretaceous seas than in the present oceans (Jenkyns 1980). Therefore, the offshore and open marine shelf environments apparently would have been rather inhospitable to the majority of these animals. This can be observed at Retiro 26, where the benthic fauna is predominantly composed only of thin-shelled, semi-infaunal inoceramids and crustaceans.

The inoceramids from Retiro 26 are allochthonous, as indicated by their broken and isolated valves; however, the distance of transport was short. Apparently, very thin-shelled inoceramids (such as *Mytiloides*, *Rhyssomytiloides* and *Sergipia*) did not have a particularly strong articulation

(Tanabe 1973), which explains the common occurrence of isolated valves without preserved articulation. For the same reason, most of the non-divergently ornamented specimens have broken margins and are preserved as internal or composite moulds.

The non-preservation of calcareous organisms in bed D, if they existed at all in the area during the time, is not easily explained. The inoceramids are commonly poorly represented in high-energy habitats (Kauffman 1967), as in bed B of the Retiro 26 sequence. It is possible that, following deposition of bed C, the high-energy conditions persisted, creating an inhospitable habitat for the pioneering colonization by inoceramids. However, the absence of ammonites makes this hypothesis less probable, suggesting a greater depth. High energy may be true for bed L, where ammonites (but not inoceramids) were found.

The development of a wedge-shaped bed between units E and F of the sequence may be explained by slumping of the sediments probably caused by a minor tilting of the basin.

The presence of post-Albian bioherm-like bodies is recorded for the first time in Sergipe. They occur in two units and indicate local, early submarine cementation. Berthou & Bengtson (1988) comment on this occurrence of bioherm-like bodies and suggest deposition at or near the carbonate platform margin, where also breccias can be formed.

The most abundant trace fossils in the Cotinguiba Formation are, according to Bengtson (1983), of *Planolites* type. However these traces were not observed in the sequence here studied, where burrows of *Thalassinoides* type predominate, particularly on erosion surfaces. Small, paired ammonite aptychi, previously not mentioned in the literature, occur throughout the sequence.

Conclusions

1. The middle and upper parts of the lower Turonian of the Retiro area in Sergipe present a rich and fairly well preserved inoceramid fauna, consisting of endemic species and cosmopolitan forms which suggest a subtropical sea.

2. The dominantly shallow marine environment of the outer continental shelf, revealed by the palaeontological associations and sedimentary structures, probably changed from a quiet and well oxygenated sea to an intermittently stirred and poorly oxygenated sea, finally returning to a well oxygenated open sea with good circulation.

3. There were bathymetric oscillations and short-lived regressive episodes during the deposition of the Cotinguiba Formation, as indicated by the ten discontinuity surfaces, some with bioturbation, observed at Retiro 26.

4. The epicontinental sea existing in Sergipe probably included a barrier separating it from the main oceanic system during the middle part of the early Turonian. Subsequently (late early Turonian), the barrier disappeared and an open sea with oceanic circulation was established.

5. Two inoceramid associations can be identified at Retiro 26, separated by a barren interval: a *Mytiloides mytiloides*

association in the middle part of the lower Turonian and a *Mytiloides hercynicus* association referred to the upper lower Turonian.

6. Among the inoceramids, species of *Mytiloides* are the most abundant (*M. labiatus* and *M. hercynicus* groups), followed by species of *Sergipia* and *Rhyssomytiloides*, respectively.

7. The occurrence of inoceramids with divergent ornamentation in the lower Turonian is unexpected and probably reflects an adaptation to a new ecological niche.

8. Phylogenetically, *Mytiloides* of the *M. labiatus* group is the probable ancestor of species of *Rhyssomytiloides*.

9. The inoceramids of Retiro 26 are allochthonous, deposited at a short distance from their life sites.

10. The time span corresponding to the deposition of the Retiro 26 sequence can be estimated at approximately 0.2–0.4 Ma. The average rate of sedimentation is in the order of 15–20 cm/1000 years, with a maximum rate of up to 30–40 cm/1000 years during periods of continuous deposition.

Summary

The lithology and palaeontology of the Retiro 26 sequence of the Sergipe Basin in northeastern Brazil is described, with particular emphasis on the occurrence of inoceramids. Study of the 35 m exposed of the Cotinguiba Limestone Formation (Laranjeiras facies) has revealed an abundant and fairly well preserved lower Turonian inoceramid assemblage, associated with ammonites, crustaceans, echinoids, algae and fish remains. The fossils and the lithology suggest during this time a shallow sea pertaining to the outer carbonate platform, with oscillating water depth, occasionally with strong submarine currents and ephemeral reductions of the oxygen levels at the sea bottom. The inoceramids are partly endemic, perhaps owing to the presence of a dispersal barrier subparallel to the coast, which separated the shallow sea from the open South Atlantic Ocean during the middle part of the early Turonian. This barrier disappeared during the later part of the early Turonian and an open sea circulation was established, with a subsequent increase in the number of cosmopolitan inoceramids species. The South Temperate Subprovince, to which this fauna belonged, was a region of subtropical waters. The sedimentation at Retiro 26 included breaks, indicated by breccia beds, paraconformities and erosion surfaces with bioturbation. Two successive inoceramid associations are recognized: a *Mytiloides mytiloides* assssociation (corresponding to the middle part of the lower Turonian) and a *Mytiloides hercynicus* association (corresponding to the upper part of the lower Turonian). Inoceramid species with divergent ornamentation are described under the new genus *Rhyssomytiloides*: *R. mauryae* (Hessel), *R.* aff. *mauryae*, *R. alatus* (Hessel), *R. bengtsoni* n. sp., *R. beurleni* n. sp., and *R. retirensis* n. sp. Other species described are: *Mytiloides modeliaensis* (Sornay), *M. mytiloides* (Mantell), *M. submytiloides?* (Seitz), *M. transiens* (Seitz), *M. hercynicus* (Petrascheck), *M.*
aff. *goppelnensis* (Badillet & Sornay), *Sergipia* aff. *posidonomyaformis* (Maury) and *S. hartti* n. sp.

Resumo

A seqüência litológica e paleontológica da localidade de Retiro 26 no Nordeste do Brasil é descrita e comentada com ênfase na ocorrência de inoceramídeos. O estudo de 35 m de calcários da formação Cotinguiba (fácies Laranjeiras) revelou uma abundante e relativamente bem preservada fauna de inoceramídeos no Turoniano inferior da bacia de Sergipe, ocorrendo juntamente com amonóides, equinóides, crustáceos, peixes e algas. Os fósseis e a litologia observados em Retiro 26 sugerem que durante o referido período havia um mar raso sobre uma plataforma carbonática externa, com batimetria oscilante, ocasionalmente com fortes correntes submarinas e efêmeras reduções dos níveis de oxigênio junto ao substrato. Os inoceramídeos são parcialmente endêmicos provavelmente devido a uma barreira de dispersão subparalela à costa que separava um mar epicontinental do sistema oceânico do Atlântico Sul durante a parte média do Turoniano inferior. No final do Turoniano inferior esta barreira deve ter desaparecido e a circulação de mares abertos pode ser estabelecida, com o aumento do número de espécies de inoceramídeos cosmopolitas. A Subprovíncia Temperada Sul, da qual a fauna estudada faz parte, apresentava então águas subtropicais. A sedimentação em Retiro 26 mostra diversas inconformidades, evidenciadas por níveis brechóides, paraconformidades ou por superfícies de erosão bioturbadas. Duas associações de inoceramídeos são reconhecidas na seqüência: a de *Mytiloides mytiloides* (correspondente à parte média do Turoniano inferior) e a de *Mytiloides hercynicus* (correspondente à parte superior do Turoniano inferior). Formas de inoceramídeos divergentemente ondulados são descritas sob o novo gênero *Rhyssomytiloides*: *R. mauryae* (Hessel), *R. alatus* (Hessel), *R.* aff. *mauryae*, *R. bengtsoni* n. sp., *R. beurleni* n. sp. e *R. retirensis* n. sp. Outras espécies ocorrentes em Sergipe e aqui descritas são: *Mytiloides modeliaensis* (Sornay), *M. mytiloides* (Mantell), *M. submytiloides?* (Seitz), *M. transiens* (Seitz), *M. hercynicus* (Petrascheck), *M.* aff. *goppelnensis* (Badillet & Sornay), *Sergipia* aff. *posidonomyaformis* e *S. hartti* n. sp.

Résumé

La séquence lithologique et paléontologique de Retiro 26, localité du nord-est du Brésil, est décrite et commentée. Une attention particulière a été donnée aux Inocérames. L'étude de 35 m de calcaires de la Formation Cotinguiba (Faciès Laranjeiras) a revèlé un assemblage abondant et assez bien conservé d'Inocérames du Turonien inférieur du bassin de Sergipe; il est accompagné d'Ammonites, d'Échinides, de Crustacés, de poissons et d'algues. Les organismes et la lithologie observés à Retiro 26 indiquent l'existence d'une mer peu profonde, à bathymétrie assez variable, avec des courants marins parfois forts et avec d'occasionnels moments de réduction d'oxygène dans les fonds marins. Un endémisme partiel est constaté pour la faune d'Inocérames; il est peut-être du à une barrière de dispersion sub-parallèle à la côte qui séparait, au Turonien inférieur moyen, une mer intérieure de l'Océan Atlantique Sud. Cette barrière n'existait plus à la fin du Turonien inférieur et une circulation de mer ouverte s'est établie dans cette région. Les Inocérames de cette période étaient des espèces cosmopolites. La Sous-province 'Sud Tempéré' à laquelle cette faune appartenait était subtropicale. La sédimentation à Retiro 26 a formé divers couches brechoïdes, des paraconcordances et des surfaces d'érosion avec bioturbations. On peut distinguer deux associations d'Inocérames: dans la partie moyènne du Turonien inférieur, l'association à *Mytiloides mytiloides*, et dans la partie supérieur du Turonien inférieur, celle à *Mytiloides hercynicus*. Les Inocérames à ornementation divergente sont placés dans un nouveau genre *Rhyssomytiloides*: *R. mauryae* (Hessel), *R. alatus* (Hessel), *R.* aff. *mauryae*, *R. bengtsoni* n. sp., *R. beurleni* n. sp. et *R. retirensis* n. sp. Les autres espèces reconnues dans les dépôts turoniens du Sergipe sont les suivantes: *Mytiloides modeliaensis* (Sornay), *M. mytiloides* (Mantell), *M. submytiloides*? (Seitz), *M. transiens* (Seitz), *M. hercynicus* (Petrascheck), *M.* aff. *goppelnensis* (Badillet & Sornay), *Sergipia* aff. *posidonomyaformis* et *S. hartti* n. sp.

References

Asmus, H.E. 1981: Geologia das bacias marginais atlânticas meso-zóicas–cenozóicas do Brasil. *In* W. Volkheimer & E.A. Musacchio (eds.): *Cuencas sedimentarias del Jurásico y Cretácico de America del Sur 1*, 127–155. Comité Sudamericano del Jurásico y Cretácico, Buenos Aires.

Asmus, H.E. 1982: Significado geotectônico das feições estruturais das bacias brasileiras e áreas adjacentes. *Anais do XXXII Congresso Brasileiro de Geologia* [Salvador, 1982] *4*, 1547–1557. Sociedade Brasileira de Geologia, São Paulo.

Asmus, H.E. & Campos, D. de A. 1983: Stratigraphic division of the Brazilian continental margin and its paleogeographic significance. *Zitteliana 10*, 265–276. München.

Asmus, H.E. & Guazelli, W. 1981: Descrição sumária das estruturas da margem continental brasileira e das áreas oceânicas e continentais adjacentes: hipóteses sobre o tectonismo causador e implicações para os prognósticos do potencial de recursos minerais. *Petrobrás, CENPES, DINTEP, Projeto REMAC 9*, 187–269. Rio de Janeiro.

Asmus, H.E. & Porto, R. 1980: Diferenças nos estágios iniciais de evolução da margem continental brasileira: possíveis causas e implicações. *Anais do XXXI Congresso Brasileiro de Geologia* [Camboriú] *1*, 225–239. Sociedade Brasileira de Geologia, São Paulo.

Aurich, N., Schaller, H. & Barros, M.C. 1972: Guaricema – primeiro campo de petróleo na plataforma continental brasileira. *Anais do XXV Congresso Brasileiro de Geologia* [São Paulo, 1971] *3*, 253–262. Sociedade Brasileira de Geologia, São Paulo.

Ayyasami, K. & Banerji, R.K. 1984: Cenomanian–Turonian transition in the Cretaceous of southern India. *Bulletin of the Geological Society of Denmark 33(1–2)*, 21–30. Copenhagen.

Badillet, G. & Sornay, J. 1980: Sur quelques formes du groupe d'*Inoceramus labiatus* décrites par O.Seitz. Impossibilité d'utiliser ce groupe pour une datation stratigraphique du Turonien inférieur du Saumurois (France). *Comptes Rendus de l'Académie des Sciences 290*, 323–325. Paris.

Bandeira Jr, A.N. 1978: Sedimentologia e microfácies calcárias das Formações Riachuelo e Cotinguiba da Bacia Sergipe/Alagoas. *Boletim Técnico da Petrobrás 21(1)*, 17–69. Rio de Janeiro.

Basse, É. 1934: Étude géologique du Sud-Ouest de Madagascar. *Mémoires de la Société Géologique de France, Série A 24*, 5–157, Pls. A–L + Pls. I–III. Paris.

Bender, F. 1959: Zur Geologie des Küsten-Beckens von Sergipe, Brasilien. *Geologisches Jahrbuch 77*, 1–33, 1 map. Hannover.

Bengtson, P. 1979: A bioestratigrafia esquecida: avaliação dos métodos bioestratigráficos no Cretáceo médio do Brasil. *Anais da Academia Brasileira de Ciências 51(3)*, 535–544. Rio de Janeiro.

Bengtson, P. 1983: The Cenomanian–Coniacian of the Sergipe Basin, Brazil. *Fossils and Strata 12*, 1–78. Oslo.

Bengtson, P. 1986: Brazil. *In* R.A. Reyment & P. Bengtson (comp.): *Events of the mid-Cretaceous: final report on results obtained by IGCP Project No. 58, 1974–1985*, 114–121. (Physics and Chemistry of the Earth 16.) Pergamon Press. Oxford.

Bengtson, P. & Berthou, P.Y. 1983: Microfossiles et Échinodermes *incertae sedis* des dépôts albiens a coniaciens du bassin de Sergipe-Alagoas, Brésil. *Cahiers de Micropaléontologie 1982(3)* [for 1982], 13–23. Paris.

Berthou, P.Y. & Bengtson, P. (1988): Stratigraphic correlation by microfacies of the Cenomanian–Coniacian of the Sergipe Basin, Brazil. *Fossils and Strata 21*, 1–88, Pls. 1–48. Oslo.

Berthou, P.Y. & Lauverjat, J. 1978: Le bassin occidental portugais de l'Albien au Campanien. *Annales du Muséum d'Histoire Naturelle de Nice 4*, I.1–I.14 [Mid-Cretaceous Events, Uppsala 1975 – Nice 1976], Nice.

Beurlen, G. 1970: Uma nova fauna de amonóides da Formação Sapucari/Laranjeiras (Cretáceo de Sergipe): considerações sobre sua bioestratigrafia. *Boletim Técnico da Petrobrás 12(2)* [for 1969], 147–169. Rio de Janeiro.

Beurlen, G. 1982: Bioestratigrafia e geoistória da seção marinha da margem continental brasileira. *Boletim Técnico da Petrobrás 25(2)*, 77–83. Rio de Janeiro.

Beurlen, K. 1961: O Turoniano marinho no nordeste do Brasil. *Boletim da Sociedade Brasileira de Geologia 10(2)*, 39–52. São Paulo.

Beurlen, K. 1964: *A fauna do Calcário Jandaíra da região de Mossoró (Rio Grande do Norte)*. 215 pp. Ed. Pongetti, Rio de Janeiro (Coleção Mossoroense, C 13).

Beurlen, K. 1967a: Estratigrafia da faixa sedimentar costeira Recife–João Pessoa. *Boletim da Sociedade Brasileira de Geologia 16(1)*, 43–53. São Paulo.

Beurlen, K. 1967b: Paleontologia da faixa costeira Recife–João Pessoa. *Boletim da Sociedade Brasileira de Geologia 16(1)*, 73–79. São Paulo.

Böse, E. 1923: Algumas faunas cretácicas de Zacatecas, Durango y Guerrero. *Boletín del Instituto de Geología de México 42*, 1–219, Pls. 1–19. México.

Böse, E. & Cavins, O.A. 1928: The Cretaceous and Tertiary of southern Texas and northern Mexico. *University of Texas Bulletin 2748*, [for 1927], 7–142. Austin.

Brongniart. A. 1822: Sur quelques terrains de craie hors du bassin de Paris. *In* G. Cuvier & A. Brongniart: *Description géologique des environs de Paris.*, 80–101, 382–400, Pls. 3–9.G. Dufour et E. D'Ocagne, Paris.

Cooper, M.R. 1978: The mid-Cretaceous (Albian–Turonian) biostratigraphy of Angola. *Annales du Muséum d'Histoire Naturelle de Nice 4*, XIV.1–XIV.22 [Mid-Cretaceous Events, Uppsala 1975 – Nice 1976], Nice.

Cox, L.R. 1969: Family Inoceramidae Giebel, 1852. *In* R.C. Moore (ed.): *Treatise on Invertebrate Paleontology: Pt. N1. Mollusca 6 – Bivalvia*, N314–N321. Geological Society of America and University of Kansas Press, Boulder, Lawrence.

Crame, J.A. 1981: Upper Cretaceous inoceramids (Bivalvia) from the James Ross Island Group and their stratigraphical significance. *British Antarctic Survey Bulletin 53*, 29–56. London.

Crame. J.A. 1982: Late Mesozoic bivalve biostratigraphy of the Antarctic Peninsula region. *Journal of the Geological Society of London 139*, 771–778. London.

Dartevelle, E. & Freneix, S. 1957: Mollusques fossiles du Crétacé de la côte occidentale d'Afrique du Cameroun à l'Angola. II – Lamellibranches. *Annales du Muséum Royale du Congo Belge 20(8)*, 1–274. Tervuren.

Dhondt, A.V. 1983: Campanian and Maastrichtian Inoceramids: a review. *Zitteliana 10*, 689–701. München.

Duarte, A.G. 1936: Petróleo e condições para sua ocorrência no Estado de Sergipe. *Mineração e Metalurgia 1(3)*, 116–117. Rio de Janeiro.

Duarte, A.G. 1938: Estado de Sergipe. *In* E.P. de Oliveira: *Relatório annual do Director, anno de 1935*, 39–46. Departamento Nacional da Produção Mineral, Serviço Geológico e Mineralógico, Rio de Janeiro.

Etayo-Serna, F. 1964: Posición de las faunas en los depósitos cretácicos colombianos y su valor en la subdivisión cronológica de los mismos. *Boletín de Geología, Universidad Industrial de Santander 16–17,* 1–142. Bucaramanga.

[Freitas, L.C. da S. 1984: *Nanofósseis calcários e sua distribuição (Aptiano–Mioceno) na Bacia Sergipe/Alagoas.* 247 pp., Dissertação de Mestrado, Departamento de Geociências da Universidade Federal do Rio de Janeiro, Rio de Janeiro. Unpublished M.Sc. Thesis.]

Glazunov, V.S. 1972: Novye pozdnemelovye inotseramidy Sakhalina. [New Late Cretaceous inoceramids from Sakhalin.] *In Novye vidy drevnikh rastenij i bespozvonochnykh SSSR,* 123–131, Pls. 34–36. Moskva.

Hallam, A. 1981: *Facies interpretation and the stratigraphic record.* 291 pp. W.H. Freeman & Co, Oxford.

Hallam, A. *et al.* 1985: Jurassic to Paleogene: Part I. Jurassic and Cretaceous geochronology and Jurassic to Paleogene magnetostratigraphy, *Geological Society of London, Memoir 10,* 118–140. London.

Haq, B.U., Hardenbol, J. & Vail, P.R. 1987: Chronology of fluctuating sea levels since the Triassic. *Science 235.* 1156–1167. New Haven.

Harland, W.B. *et al.* 1982: *A Geologic Time Scale.* 131 pp. Cambridge University Press, Cambridge.

Hartt, C.F. 1868: A naturalist in Brazil. *The American Naturalist 2(1),* 1–93. Salem.

Hartt, C.F. 1870: *Geology and Physical Geography of Brazil.* xxi + 620 pp. Fields, Osgood & Co, Boston. [Reprint in 1975 by Robert E. Krieger Publ. Co, New York.]

Hattin, D.E. 1962: Stratigraphy of the Carlile Shale (Upper Cretaceous) in Kansas. *Bulletin of the State Geological Survey of Kansas 156,* 1–155. Lawrence.

Hayami, I. 1975: A systematic survey of the Mesozoic Bivalvia from Japan. *Bulletin of the University Museum, The University of Tokyo 10,* 1–249. Tokyo.

Heinz, R. 1928a: Über die Oberkreide-Inoceramen Süd-Amerikas und ihre Beziehungen zu denen Europas und anderer Gebiete. *Mitteilungen aus dem Mineralogisch-Geologischen Staatsinstitut in Hamburg 10,* 41–97, Pls. 4–6. Hamburg.

Heinz, R. 1928b: Über die bisher wenig beachtete Skulptur der Inoceramen-Schale und ihre stratigraphische Bedeutung. *Mitteilungen aus dem Mineralogisch-Geologischen Staatsinstitut in Hamburg 10,* 1–39, Pls.1–3. Hamburg.

Heinz, R. 1932: Sur les Inocérames de Madagascar. *Annales Géologiques du Service de Mines 2,* 55–61. Tananarive.

Heinz, R. 1933: Inoceramen von Madagaskar und ihre Bedeutung für die Kreide-Stratigraphie. *Zeitschrift der Deutschen Geologischen Gesellschaft 85(4),* 241–259, Pls. 16–22. Hannover.

Herngreen, G.F.W. 1975: Palynology of middle and upper Cretaceous strata in Brazil. *Mededelingen Rijks Geologische Dienst, Nieuwe Serie 26(3),* 39–91. Haarlem.

Hessel, M.H.R. 1986: Alguns inoceramídeos (Bivalvia) radialmente ondulados do Turoniano inferior de Sergipe. *Departamento Nacional da Produção Mineral, Série Geologia 27,* 227–237. [Coletânea de Trabalhos Paleontológicos, Série Didática 2.] Brasília.

Jenkyns, H.C. 1980: Cretaceous anoxic events: from continents to ocean. *Journal of the Geological Society 137(2),* 171–188. London.

Juignet, P & Kennedy, W.J. 1974: Structures sédimentaires et mode d'accumulation de la craie du Turonien supérieur et du Sénonien du Pays de Caux. *Bulletin du B.R.G.M., Deuxième série, 4(1),* 19–47. Paris.

Kauffman, E.G. 1967: Coloradoan macroinvertebrate assemblages – central Western Interior, United States. *Paleo-environments of the Cretaceous Seaway – a Symposium,* 67–143. Boulder.

Kauffman, E.G. 1969: Form, function and evolution. *In* R.C. Moore (ed.): *Treatise on Invertebrate Paleontology: Pt. N 1 Mollusca 6 – Bivalvia,* N129–N205. Geological Society of America and University of Kansas Press, Boulder, Lawrence.

Kauffman, E.G. 1973: Cretaceous Bivalvia. *In* A. Hallam (ed.): *Atlas of Palaeobiogeography,* 353–383. Elsevier, Amsterdam.

Kauffman, E.G. 1975: Dispersal and biostratigraphic potential of Cretaceous benthonic Bivalvia in the Western Interior. *In* W.G.E. Caldwell (ed.): The Cretaceous System in the Western Interior of North America. *Geological Association of Canada, Special Paper 13,* 163–194. Waterloo, Ontario.

Kauffman, E.G. 1977a: Evolutionary rates and biostratigraphy. *In* E.G. Kauffmann & J.E. Hazel (eds.): *Concepts and Methods of Bio-*

stratigraphy, 109–141. Dowden, Hutchinson & Ross, Stroudsburg.

Kauffman, E.G. 1977b: Systematic, biostratigraphic and biogeographic relationships between middle Cretaceous Euramerican and North Pacific Inoceramidae. *Palaeontological Society of Japan, Special Papers 21,* 169–212. Fukuoka.

Kauffman, E.G. 1978a: Middle Cretaceous bivalve zones and stage implications in the Antillean Subprovince, Caribbean Province. *Annales du Muséum d'Histoire Naturelle de Nice 4,* XXX.1–XXX.11 [Mid-Cretaceous Events, Uppsala 1975 – Nice 1976], Nice.

Kauffman, E.G. 1978b: British middle Cretaceous inoceramid biostratigraphy. *Annales du Muséum d'Histoire Naturelle de Nice 4,* IV.1–IV.12 [Mid-Cretaceous Events, Uppsala 1975 – Nice 1976], Nice.

Kauffman, E.G. 1978c: An outline of middle Cretaceous marine history and inoceramid biostratigraphy in the Bohemian Basin, Czechoslovakia. *Annales du Muséum d'Histoire Naturelle de Nice 4,* XIII.1–XIII.12 [Mid-Cretaceous Events, Uppsala 1975 – Nice 1976], Nice.

Kauffman, E.G. 1979: Cretaceous. *In* R.A. Robinson & C. Teichert (eds.): *Treatise on Invertebrate Paleontology: Pt. A Introduction,* A418–A487. The Geological Society of America and The University of Kansas, Boulder, Lawrence.

Kauffman, E.G., Hattin, D.E. & Powell, J.D. 1977: Stratigraphic, paleontologic and paleoenvironmental analysis of the Upper Cretaceous rocks of Cimarron County, northwestern Oklahoma. *Memoires of the Geological Society of America 149,* 1–150. Boulder.

Kauffman, E.G. & Sohl, N. 1978: Middle Cretaceous events in the Caribbean Province. *Annales du Muséum d'Histoire Naturelle de Nice 4,* XXIX.1–XXIX.5 [Mid-Cretaceous Events, Uppsala 1975 – Nice 1976], Nice.

Kauffman, E.G., Cobban, W.A. & Eicher, D.L. 1978: Albian through lower Coniacian strata, biostratigraphy and principal events, Western Interior United States. *Annales du Muséum d'Histoire Naturelle de Nice 4,* XXIII.1–XXIII.52 [Mid-Cretaceous Events, Uppsala 1975 – Nice 1976], Nice.

Kauffman, E.G. & Bengtson, P. 1985: Mid-Cretaceous inoceramids from Sergipe, Brazil: a progress report. *Cretaceous Research 6,* 311–315. London.

Keller, S. 1982: Die Oberkreide der Sack-Mulde bei Alfeld (Cenoman–Unter-Coniac): Lithologie, Biostratigraphie und Inoceramen. *Geologisches Jahrbuch, Serie A 64,* 1–171. Hannover.

Kennedy, W.J. & Garrison, R.E. 1975: Morphology and genesis of nodular chalks and hardgrounds in the Upper Cretaceous of southern England. *Sedimentology 22(3),* 311–386. Oxford.

Kennedy, W.J. & Klinger, H.C. 1972: Hiatus concretions and hardgrounds horizons in the Cretaceous of Zululand, South Africa. *Palaeontology 15,* 539–549, Pls. 106–108. London.

Kennedy, W.J. & Klinger, H.C. 1975: Cretaceous faunas from Zululand and Natal, South Africa: Introduction, stratigraphy. *Bulletin of the British Museum (Natural History), Geology 25(4),* 265–315, 1 pl. London.

Klinger, H.C. 1977: Cretaceous deposits near Bogenfels, South West Africa. *Annals of the South African Museum 73(3),* 81–92. Cape Town.

[Lana, M. da C. 1985: *Rifteamento na Bacia de Sergipe/Alagoas, Brasil.* 124 pp. Dissertação de Mestrado, Universidade Federal de Ouro Preto, Ouro Preto. Unpublished M.Sc. Thesis.]

Lima, M.R. de & Boltenhagen, E. 1981: Estudo comparativo da evolução das microfloras afro-sul-americanas: II. O Cretáceo superior. *In* Y.T. Sanguinetti (ed.): *Anais do II Congresso Latino-Americano de Paleontologia 1,* 373–383. Porto Alegre.

Lupu, M. 1978: Preliminary report on Albian–Turonian deposits in Romania. *Annales du Muséum d'Histoire Naturelle de Nice 4,* XIV.1–XIV.18 [Mid-Cretaceous Events, Uppsala 1975 – Nice 1976], Nice.

Mantell, G. 1822: *Fossils of the South Downs; or illustrations of the geology of Sussex,* 327 pp. Lupton Relfe, London.

Marks, J.G. 1956: Pacific Coast geological province. *In* W.F. Jenks (ed.): Handbook on South American geology. *Memoirs of the Geological Society of America 65,* 277–288. New York.

Matsumoto, T. 1959: Zoning of the Upper Cretaceous in Japan and adjacent areas with special reference to world-wide correlation. *Congreso Geológico Internacional, XX Sesión* [México 1956], 347–381. Symposium del Cretácico, México.

Matsumoto, T. & Noda, M. 1975: Notes on *Inoceramus labiatus*

(Cretaceous Bivalvia) from Hokkaido. *Transactions and Proceedings of the Palaeontological Society of Japan, New Series 100*, 188–208, Pl. 18. Tokyo.

Matsumoto, T. & Noda, M. 1985: A new inoceramid (Bivalvia) species from the upper Campanian (Cretaceous) of Hokkaido. *Proceedings of the Japan Academy, Ser. B 61(1)*, 9–11. Tokyo.

Maury, C.J. 1925: Fosseis terciarios do Brasil, com descrição de novas formas cretaceas. *Serviço Geologico e Mineralogico do Brasil, Monographia 4*, 1–705, Pls. 1–24, 1 map, 2 tables. Rio de Janeiro.

Maury, C.J. 1937: O Cretaceo de Sergipe. *Serviço Geologico e Mineralogico do Brasil, Monographia 11*, 1–283, 6 tables; *Album das Estampas*: I–XXXV, Pls. 1–28. Rio de Janeiro.

Morais Rego, L.F. de 1933: Notas sobre a geologia, a geomorfologia e os recursos minerais de Sergipe. *Anais da Escola de Minas (de Ouro Preto) 24*, 31–84. Ouro Preto.

Nicol, D. 1964: An essay on size of marine pelecypods. *Journal of Paleontology 38(5)*, 968–974. Tulsa.

Noda, M. 1983: Notes on the so-called *Inoceramus japonicus* (Bivalvia) from the Upper Cretaceous of Japan. *Transactions and Proceedings of the Palaeontological Society of Japan, New Series 132*, 191–219, Pls. 41–46. Tokyo.

Noguti, I. & Santos, J.F. 1973: Zoneamento preliminar por foraminíferos planctônicos do Aptiano ao Mioceano na plataforma continental do Brasil. *Boletim Técnico da Petrobrás 15(3)* [for 1972], 265–283. Rio de Janeiro.

Odin, G.S. 1985: Concerning the numerical ages proposed for the Jurassic and Cretaceous geochronology. *Geological Society of London, Memoir 10*, 196–198. London.

Offodile, M.E. 1976: The Geology of the Middle Benue, Nigeria. *Publications from the Palaeontological Institution of the University of Uppsala, Special Volume 4*, 1–166. Uppsala.

Offodile, M.E. & Reyment, R.A. 1977: Stratigraphy of the Keana-Awe area of the middle Benue region of Nigeria. *Bulletin of the Geological Institutions of the University of Uppsala, New Series 7*, 37–66. Uppsala.

Ojeda, H.A.O. 1983: Estrutura, estratigrafia e evolução das bacias marginais brasileiras. *Revista Brasileira de Geociências 11(4)* [for 1981], 257–273. São Paulo.

Ojeda, H.A.O. 1984: Estrutura e evolução das bacias mesozóicas emersas da margem continental brasileira. *Revista Brasileira de Geociências 13(2)* [for 1983], 71–83. São Paulo.

Ojeda, H.A.O. & Fugita, A.M. 1976: Bacia Sergipe/Alagoas: geologia regional e perspectivas petrolíferas. *Anais do XXVIII Congresso Brasileiro de Geologia* [Porto Alegre, 1974] *1*, 137–158. Sociedade Brasileira de Geologia, São Paulo.

Oliveira, P.E. de 1940: Idade do calcário de Calumbi (Sergipe). *Notas Preliminares e Estudos 19*, 1–12, Pls. 1–2. Departamento Nacional da Produção Mineral, Divisão de Geologia e Mineralogia, Rio de Janeiro.

Olsson, A.A. 1956: Colombia. *In* W.F. Jenks (ed.): *Handbook on South American geology*, 293–326. *Memoirs of the Geological Society of America 65*. New York.

Pergament, M.A. & Treger [Tröger], K.A. 1979: Stratigraficheskoe znachenie radial'noj skul'ptury pozdnemelovykh inotseramov. [Stratigraphical significance of the radial sculpture in Late Cretaceous inoceramids.] *Izvestiya Akademii nauk SSSR, seriya geologicheskaya 7*, 71–79. Moskva. [Translated by Geological Survey of Canada Library (Report 460321), 1985. 17 pp. Ottawa.]

Petrascheck, W. 1904: Über Inoceramen aus der Kreide Böhmens und Sachsens. *Jahrbuch der Kaiserlich-königlichen geologischen Reichsanstalt 53(1)*, 153–168, Pls. 1–2. Wien.

Petri, S. 1962: Foraminíferos cretáceos de Sergipe. *Boletim da Faculdade de Filosofia, Ciências e Letras da Universidade de São Paulo 265 (Geologia 20)*, 1–140, Pls. 1–21. São Paulo.

Ponte, F.C., Fonseca, J.R. & Carozzi, A.V. 1980: Petroleum habitats in the Mesozoic–Cenozoic of the continental margin of Brazil. *In* A.D. Miall (ed.): *Facts and principles of world petroleum occurrence*, 857–886. Canadian Society of Petroleum Geologists, Memoir 6. Ottawa.

Raine, J.I., Speden, I.G. & Strong, C.P. 1981: New Zealand. *In* R.A. Reyment & P. Bengtson (eds.): *Aspects of Mid-Cretaceous Regional Geology*, 221–267. Academic Press, London.

Reeside Jr, J.B. 1928: Report on the fossils. *In* T. Wasson & J.H. Sinclair: Geological explorations East of the Andes in Ecuador,

1268–1281. *Bulletin of the American Association of Petroleum Geologists 11(12)*. Chicago.

Regali, M., Uesugui, N. & Santos, A.S. 1974: Palinologia dos sedimentos meso–cenozóicos do Brasil (I). *Boletim Técnico da Petrobrás 17(3)*, 177–191, Figs. 5–6. Rio de Janeiro.

Reyment, R.A. 1955: Upper Cretaceous Mollusca (Lamellibranchia and Gastropoda) from Nigeria. *Colonial Geology and Mineral Resources 5(2)*, 127–155. London.

Reyment, R.A. & Tait, E.A. 1972: Biostratigraphical dating of the early history of the South Atlantic Ocean. *Philosophical Transactions of the Royal Society of London, B – Biological Sciences 264(858)*, 55–95, Fig. 11, Pls. 3–5. London.

Reyment, R.A., Bengtson, P. & Tait, E.A. 1976: Cretaceous transgressions in Nigeria and Sergipe–Alagoas (Brazil). *Anais da Academia Brasileira de Ciências 48 (Suplemento)*, 253–264 [F.F.M. de Almeida (ed.): Continental Margins of Atlantic Type]. Rio de Janeiro.

Riedel, L. 1932: Die Oberkreide vom Mungofluss in Kamerun und ihre Fauna. *Beiträge zur geologischen Erforschung der deutschen Schutzgebiete 16*, 1–154, Pls. 1–33. Berlin.

Robaszynski, F. 1978: Approche stratigraphique du Cénomano–Turonien dans le Hainaut Franco/Belge et le Nord de la France. *Annales du Muséum d'Histoire Naturelle de Nice 4*, VIII.1–VIII.23 [Mid-Cretaceous Events, Uppsala 1975 – Nice 1976], Nice.

Robaszynski, F. (coord.) 1982: Le Turonien de la région-type: Saumurois et Touraine. Stratigraphie, biozonations, sédimentologie. *Bulletin des Centres de Recherches Exploration-Production Elf-Aquitaine 6(1)*, 119–225. Pau.

Rutsch, R.F. & Salvador, A. 1954: Mollusks from the Cogollo and La Luna Formations (Cretaceous) of the Chejendé area, western Venezuela. *Journal of Paleontology 28(4)*, 417–426. Tulsa.

Sampaio, A.V. & Northfleet, A. 1975: Estratigrafia e correlação das bacias sedimentares brasileiras. *Anais do XXVII Congresso Brasileiro de Geologia* [Aracaju, 1973] *3*, 189–206. Sociedade Brasileira de Geologia, São Paulo.

Schaller, H. (ed.) 1970: Revisão estratigráfica da Bacia de Sergipe/Alagoas. *Boletim Técnico da Petrobrás 12(1)* [for 1969], 21–86, Figs. 3, 10, 11, 16A, 30, 31. Rio de Janeiro.

Seibertz, E. 1979: Biostratigraphie im Turon des SE-Münsterlandes und Anpassung an die internationale Gliederung aufgrund von Vergleichen mit anderen Oberkreide-Gebieten. *Newsletters on Stratigraphy 8(2)*, 111–123. Berlin.

Seitz, O. 1935: Die Variabilität des *Inoceramus labiatus* v. Schlotheim. *Jahrbuch der Königlichen Preussischen geologischen Landesanstalt* [for 1934] *55(1)*: 426–474. Berlin.

Seitz, O. 1959: Vergleichende Stratigraphie der Oberkreide in Deutschland und in Nordamerika mit Hilfe der Inoceramen. *Congresso Geológico Internacional, XX Sesión* [Symposium del Cretácico], 113–130. México.

Sornay, J. 1981: Inocérames (Bivalvia) du Turonien inférieur de Colombie (Amérique du Sud). *Annales de Paléontologie (Invertébrés) 67(2)*, 135–148. Paris.

Sornay, J. 1982: Inocérames du Saumurois. *In* F. Robaszynski *et al.*: Le Turonien de la région-type: Saumurois et Touraine. Stratigraphie, biozonations, sédimentologie. *Bulletin des Centres de Recherches Exploration-Production Elf-Aquitaine 6(1)*, 138–140. Pau.

Sornay, J. 1986: Inocérames. *Association Geologique Auboise, Bulletin annuel 9*, 17–30. Anglure.

Stanley, S.M. 1970: Relation of shell form to life habits of the Bivalvia (Mollusca). *Memoirs of the Geological Society of America 125*, 1–296. Boulder.

Stoliczka, F. 1871: Cretaceous fauna of southern India: III. The Pelecypoda, with a review of all known genera of this class, fossil and recent. *Palaeontologica Indica, serie 6(3)*, 1–537, Pls. 1–50. Geological Survey of India, Calcutta.

Tanabe, K. 1973: Evolution and mode of life of *Inoceramus (Sphenoceramus) naumanni* Yokoyama emend., an upper Cretaceous bivalve. *Transactions and Proceedings of the Palaeontological Society of Japan, New Series 92*, 143–184, Pls. 27–28. Tokyo.

Toshimitsu, S. 1986: A new inoceramid (Bivalvia) species from the upper Cretaceous of Hokkaido. *Proceedings of the Japan Academy, Serie B 62(1)*, 227–230. Tokyo.

Troelsen, J.C. & Quadros, L.P. de 1971: Distribuição bioestratigráfica dos nanofósseis em sedimentos marinhos (Aptiano–Mioceno)

do Brasil. *Anais da Academia Brasileira de Ciências 43 (Suplemento)*, 577–609. Rio de Janeiro.

Tröger, K.A. 1976: Evolutionary trends of Upper Cretaceous Inocerames. *Evolutionary Biology*, 193–203. Praha.

Tröger, K.A. 1981: Zur Problemen der Biostratigraphie der Inoceramen und der Untergliederung des Cenomans und Turons in Mittel- und Osteuropa. *Newsletters on Stratigraphy 9(3)*, 139–156. Berlin.

Wellman, H.W. 1959: Divisions of the New Zealand Cretaceous. *Transactions of the Royal Society of New Zealand 87*, 99–163, Pls. 10–12. Dunedin.

Wiedmann, J. & Kauffman, E.G. 1978: Mid-Cretaceous biostratigraphy of northern Spain. *Annales du Muséum d'Histoire Naturelle de Nice 4*, III.1–III.34 [Mid-Cretaceous Events, Uppsala 1975 – Nice 1976], Nice.

Willard, B. 1966: *The Harvey Bassler Collection of Peruvian Fossils*. 253 pp., 74 pls. Lehigh University, Bethlehem.

Woods, H. 1911: A monograph of the Cretaceous Lamellibranchia of England. [Vol. II Part VII]. *Palaeontographical Society [Monograph]* [for 1910] 261–284, Pls. 45–50. London.